SpringerBriefs in Biochemistry and Molecular Biology

For further volumes:
http://www.springer.com/series/10196

Alberto J. L. Macario
Everly Conway de Macario
Francesco Cappello

The Chaperonopathies

Diseases with Defective Molecular Chaperones

An Introduction and Guide to Diseases in which
Chaperones Play an Etiologic-Pathogenic Role

 Springer

Alberto J. L. Macario
Department of Microbiology and
  Immunology, School of Medicine
University of Maryland
Baltimore, MD
USA

Istituto EuroMediterraneo di Scienza
  e Tecnologia (IEMEST)
Palermo
Italy

Institute of Marine and Environmental
Technology (IMET)
Baltimore, MD
USA

Everly Conway de Macario
Department of Microbiology and
  Immunology, School of Medicine
University of Maryland
Baltimore, MD
USA

Institute of Marine and Environmental
Technology (IMET)
Baltimore, MD
USA

Francesco Cappello
Sezione di Anatomia Umana, Dipartimento
  di Biomedicina Sperimentale
  e Neuroscienze Cliniche
Università degli Studi di Palermo
Palermo
Italy

Istituto EuroMediterraneo di Scienza
  e Tecnologia (IEMEST)
Palermo
Italy

ISSN 2211-9353   ISSN 2211-9361   (electronic)
ISBN 978-94-007-4666-4   ISBN 978-94-007-4667-1   (eBook)
DOI 10.1007/978-94-007-4667-1
Springer Dordrecht Heidelberg New York London

Library of Congress Control Number: 2013931840

Printed on acid-free paper

Springer is part of Springer Science+Business Media (www.springer.com)

This book is dedicated to medical researchers and practitioners, clinicians and pathologists across all specialties. We hope that the book will help them in the planning of clinical–pathologic investigations on chaperonopathies, and also in dealing with patients and in managing a group of diseases generally ignored until now

Preface

This book is about a number of pathological conditions, the chaperonopathies, which probably existed since remote times and have been studied for years but were grouped together in a coherent nosological category only recently.

The book is intended to be an introduction to the chaperonopathies for researchers and also for practitioners in clinical medicine and pathology. It was conceived as a guide for future research. In this regard, it opens a quarry for mining. It provides the basic information, including references to learn about chaperonopathies from those that described them originally, so research projects can be prepared and launched.

The subject matter includes chaperones with demonstrated chaperoning roles and other molecules related to them evolutionarily and/or functionally. It focuses on conditions with chaperone malfunction and associated pathologies. In this regard it is important to point out that chaperones have typical chaperoning functions pertaining to protein homeostasis (i.e., assisting in protein folding, protection from aggregation, refolding, translocation, degradation, dissolution of protein aggregates, and selective autophagy) and, in many instances, other functions more or less unrelated to the typical ones. For example, Hsp60 participates in protein folding inside mitochondria but it has, in addition, other diverse functions such as interaction with components of the innate immune system, resulting in cytokine production. The book includes examples of chaperonopathies in which the typical chaperoning functions are compromised as well as examples of conditions in which the noncanonical roles of chaperones are affected. Thus, the book centers on the chaperone molecules, which if abnormal, can cause chaperonopathies, with manifestations on protein homeostasis and/or beyond.

Acknowledgments

We thank all those (too many to mention by name) who provided support and advice, and generated information so we could produce a useful book. We hope they will not be disappointed. Our special thanks are due to Prof. Rona Giffard, M.D., Ph.D. (Stanford University School of Medicine) for her critical reading of the manuscript; and to Prof. Monte S. Willis, M.D., Ph.D. (School of Medicine, University of North Carolina, Chapel Hill) for his insightful comments on the first draft and for writing the Prologue.

This is IMET 13-103 publication, and the work was done under the umbrella of the agreement between IEMEST and IMET signed March 26, 2012.

Potomac and Baltimore, Maryland, USA   Alberto J. L. Macario
December 2012   Everly Conway de Macario
                                              Francesco Cappello

Contents

Prologue

The accumulation of misfolded proteins may be most familiar to those who understand the role of protein aggregates in the pathophysiology of neurodegenerative disorders (e.g., Alzheimer's disease). The pathophysiology of these conditions is complex as the accumulation of protein aggregates not only depends on the cell's ability to prevent protein misfolding, but also on the cell's ability to degrade these proteins by the ubiquitin proteasome system and/or autophagic machineries. Chaperones (also called heat shock proteins although not all of these are chaperones and, vice versa, not all chaperones are heat shock proteins) play a critical role not only in protein folding, but also in the degradation and autophagic removal of misfolded proteins. When chaperone activity is impaired, protein aggregates, or any of the preceding protein intermediaries, can induce cell death—a process called proteotoxicity—which is one mechanism that underlies the pathophysiology of neurodegenerative diseases. Studies aiming at elucidating the role of proteotoxicity in neurodegenerative diseases have led to the discovery that increasing heat shock protein expression can improve cognitive function in preclinical models. Increased heat shock protein expression also has the beneficial side effects of augmenting the solubility of misfolded protein aggregates and enhancing degradation of the latter, experimentally. This is a significant finding since therapies exist today that induce heat shock proteins, including exercise and specific FDA (Food and Drug Administration) approved drugs, which are able to lower the burden of misfolded polypeptides and attenuate pathological progression. Experimentally, these therapies have proven quite successful in clinical scenarios where few options are available.

The importance of chaperones does not end with these well-known neurodegenerative conditions, including Alzheimer's, Parkinson's, Huntington's, and prion diseases. Chaperones also play a predominant role in a growing number of very common pathologies, including cystic fibrosis and heart failure. For example, the most common mutation causing cystic fibrosis encodes a misfolded protein that is preferentially degraded by the ubiquitin proteasome system, resulting in a lack of the protein on the cell's surface. Small molecule compounds that correct this mutant CFTR-protein misfolding have been developed to prevent its degradation and show promise in pre-clinical and preliminary clinical studies. And the list of diseases that have misfolded proteins and their associated "proteotoxicity" underlying their pathogenesis continues to grow. Not only have misfolded proteins been identified in heart failure patients, but it has also been demonstrated that cardiac expression of misfolded oligomers alone can be responsible for the observed heart failure. It is predicted that therapies that can prevent protein misfolding in heart failure patients, including promoting an increase in chaperone expression, may offer therapies so desperately needed for this most common cause of death in the United States.

Our understanding of protein quality control and the role of molecular chaperones in the pathophysiology of many human diseases, as discussed in this book, is just beginning and represents a common theme that may make advances in one field applicable to many others. By understanding the common pathophysiology of diseases based on their relationship to chaperones allows similar conditions to be studied together, systematically, for the first time in a coherent matrix. The material presented in this book represents an outstanding foundation of knowledge for both physicians and scientists in clinical practice and translational research in their quest to develop novel therapies, diagnostics, and appropriate differential diagnoses for human illnesses due to abnormal chaperones (i.e., the chaperonopathies). It is hoped that by elucidating these maladies as a significant group of important pathological conditions, many of them serious and widespread, the larger medical community will be more aware of the importance of the field and give it the attention it largely lacks at present.

November 2012   Monte S. Willis
University of North Carolina at Chapel Hill, NC, USA

Chapter 1
Overview and Book Plan

Abstract This chapter provides an introduction to the biology and pathology of molecular chaperones, many of which are heat-shock proteins, involved in protein homeostasis and other unrelated functions. When chaperones are defective structurally and/or functionally they may cause disease. These diseases in which chaperones play an etiologic-pathogenic role are the chaperonopathies. The chapter also gives a clinical-pathological overview of chaperonopathies and guidelines for their identification and diagnosis. It briefly describes how to detect and characterize a chaperonopathy in a patient. Chaperones can be useful biomarkers for disease diagnosis and monitoring, including evaluation of prognosis and response to treatment. This and the potential of chaperones for therapy, i.e., chaperonotherapy, are aspects of chaperonology also outlined in the chapter.

Keywords Heat shock protein · Molecular chaperone · Protein homeostasis · Abnormal chaperones · Chaperonopathies · Molecular chaperone disorders · Pathological chaperones · Hsp diseases · Abnormal Hsp · Hsp pathology · Hsp biomarkers

1.1 Hsp and Molecular Chaperones

A heat-shock protein (Hsp) is, strictly speaking, the product of a gene inducible by a sudden and short-lasting temperature elevation, i.e., heat shock. However, the term Hsp is used with great flexibility to indicate proteins that are the product of genes that can respond to a variety of stressors, not just heat shock. Many Hsp have chaperoning ability, in the sense that they assist nascent polypeptides to fold correctly. Other canonical functions of chaperones include, in addition to protein folding, assisting protein refolding and translocation through membranes, ushering proteins damaged beyond repair to degradation, and dissolution of protein aggregates. One may add selective autophagy. In summary, the canonical functions of chaperones pertain to protein quality control, essentially the maintenance of protein homeostasis. In addition, chaperones have other function unrelated to protein homeostasis, as it will be explained later.

A. J. L. Macario et al., *The Chaperonopathies*, SpringerBriefs in Biochemistry and Molecular Biology, DOI: 10.1007/978-94-007-4667-1_1, © The Author(s) 2013

There are several Hsp, mostly well known by physicians and pathologists, such as the small heat-shock proteins (sHsp), Hsp40, Hsp60, Hsp70, Hsp90, etc. named in consideration of their molecular weights. It has to be borne in mind that not all Hsp are chaperones and that not all chaperones are Hsp. However, the indiscriminant use of the names Hsp and chaperone in the literature has made a distinction between them almost impossible or at least impractical. Therefore, in this book, we will use both terms interchangeably.

The term chaperonin is applied to the chaperones with a molecular weight in the range 55–64 kDa and that have been classified into Group I and II. The members of each group are evolutionarily related: those of Group I are present in bacteria (in which Hsp60 is also named GroEL, or Cpn60) and in the mitochondria (Hsp60, also named Cpn60) of eukaryotes, including humans, whereas those of Group II occur in the cytoplasm of archaea (thermosome subunits) and in the cytosol of the eukaryotic cell (CCT subunits).

To simplify matters we will use the name Hsp60 for the bacterial GroEL and the mitochondrial Cpn60 throughout this book.

1.2 The Pathology of the Molecular Chaperones

Molecular chaperones (chaperones in short) are important components of all living cells in which they play a variety of roles, not just maintenance of protein homeostasis. A malfunctioning chaperone or the lack of its function may be cause of disease, a chaperonopathy. A chaperonopathy can be due to a structural defect in the chaperone molecule, or to a mechanism in which the affected chaperone is structurally normal but does not work, or works abnormally, or works for a cell that causes disease. In all these situations, the affected chaperone can be considered an, or *the*, etiological-pathogenic factor necessary to cause a pathologic condition. In this book, we will present diseases that can be considered chaperonopathies. We will also include conditions in which the etiologic-pathogenic role of chaperones has not yet been definitively demonstrated but is likely; future research will determine whether or not these conditions are genuine chaperonopathies. In addition, it is important to remember that many diseases are the result of convergence of more than one etiologic-pathogenic factor and chaperonopathies are no exception.

1.3 Scope

This book is about molecular chaperone genes-proteins and, to a limited extent, their co-chaperones and co-factors. We will not deal with other molecules, e.g., RNA and chemical compounds, sometimes also named chaperones. Examples of chemical chaperones are those also called pharmacological chaperones and that are utilized to restore a functional conformation to certain defective proteins, mostly enzymes (e.g., to treat inherited metabolic storage diseases).

This book focuses on diseases in which chaperones are etiologic-pathogenic factors, but not on diseases due to protein misfolding. The latter are due to mutations and other abnormalities of the sick protein itself. Thus, chaperonopathies and protein misfolding diseases are two distinct nosological groups, although it has to be borne in mind that some chaperonopathies may be at the basis of syndromes in which protein misfolding is one of the pathologic features.

Also included in the book are examples of pathologies associated with molecules such as McKusic-Kaufman syndrome (MKKS) and Bardet-Biedl syndrome (BBS) proteins, for which the typical, canonical function of chaperones, i.e., assisting the correct folding of nascent polypeptides, has not been demonstrated but, nonetheless, they are evolutionary related to chaperones and perform chaperoning functions in the assembly of the BBSome, a structure involved in ciliogenesis. Likewise, pathological conditions are discussed in which the implicated chaperone (e.g., Hsp60) exercises functions (e.g., Hsp60 interaction with neutrophils and other cells relevant to innate immunity) unrelated to its canonical, typical role in polypeptide folding and protein homeostasis.

The book deals with pathological conditions in which the association of disease with abnormal chaperones is firmly established. In addition, the book includes various conditions that are candidates for acceptance into the group of authentic chaperonopathies and that are still under scrutiny. The aim and the hope are that this book will stimulate research to elucidate whether or not these candidates are true chaperonopathies, namely to establish whether or not quantitative and/or qualitative abnormalities of chaperones play a pathogenic role. An illustrative example of this situation would be torsion dystonia, in which Torsin A appears to be involved. Data suggest that mutations in Torsin A (a chaperone) contribute to the pathogenesis of torsion dystonia. However, penetrance is low and expression variable. Typically, dystonia is an early onset disease when severe, but if not, namely when it starts in the 20 s for instance, the phenotype is milder, and if the carrier reaches the age of 30 without disease manifestations, it is likely that the disorder will never become clinically apparent. Is the mutated protein involved in pathogenesis? And if so, to what extent and how does it, and what cellular processes are affected by it? What other factors are involved in addition to the mutated Torsin molecule that would explain disease variability? This would be a typical situation in which investigations aimed at determining the role of a structurally abnormal chaperone in pathogenesis will significantly contribute to the understanding of a disease and, consequently, to its treatment.

We have attempted to provide a variety of examples of chaperonopathies, or candidate chaperonopathies, based on our own experience and on what is in the public domain (publications such as journals, books, and databases) but we do not claim that this book is a comprehensive review of the literature. Only illustrative examples of pertinent pathologic conditions and publications are provided. We apologize to the very many authors who have contributed substantially to the field but whose work we cannot discuss due to space limitations and to the characteristics of this book.

1.4 Clinical and Pathological Overview

The symptoms and signs of chaperonopathies are varied from almost undetectable to major complaints accompanied by morphological, biochemical, and anatomo-pathological abnormalities. For example, most of the hereditary chaperonopathies studied thus far tend to appear early in life and affect various tissues and organs, such as the central and peripheral nervous systems (CNS and PNS, respectively) and the cardiocirculatory apparatus. Acquired chaperonopathies, like those occurring in ageing individuals, may display a variety of signs and symptoms that are difficult to associate with any specific tissue or organ, but that constitute the basis of general discomfort. In addition, the clinical picture of chaperonopathies overlaps with those of diseases that currently are not identified as chaperonopathies but as other conditions such as atherosclerosis, arthritis, colitis, bronchitis, ataxias, distal neuropathies, neurodegenerative disorders, cardiomyopathies, brittle bone disease, etc. Furthermore, the chaperonopathies by mistake or collaborationism, such as certain types of cancer, are characterized by the clinical picture attributed to the tumor-caused disorders.

Hereditary disorders of the PNS are classified into motor and sensory, sensory, motor, and sensory and autonomic neuropathies. Signs and symptoms vary depending on the type and include numbness, tingling, and foot and hand pain in the sensory neuropathies. The motor neuropathies show sarcopenia and weakness in the lower portion of the legs and feet. When the autonomic nerves are affected abnormal sweating, postural hypotension, and insensitivity to pain, are components of the clinical picture. In addition, when muscular atrophy is present all the signs and symptoms typical of such condition are evident, such as foot, calf and other body deformities, including scoliosis. Since these diseases are mostly hereditary, similar signs and symptoms are sometimes present in the patient's relatives, although they may be affected in various degrees that are not necessarily the same in all of them. Clinical pathology work-up that helps in the diagnosis include blood and genetic tests, nerve conduction determination, and analysis of nerve and muscle biopsies.

An example of severe chaperonopathy is MitCHAP-60, a disease associated with a mutation in the mitochondrial chaperonin of Group I Hsp60, or Cpn60 (see Sect. 4.2). It is an autosomal-recessive inherited neurodegenerative disease, with early onset, and severe brain lesions that can be fatal. Clinically, there is spastic paraplegia with increased tendon reflexes including Babinski, no head control, truncal hypotonia, strabismus, and nystagmus, all accompanied by psychomotor retardation. Hypomyelinating leukodystrophy is evident at magnetic resonance imaging (MRI).

MitCHAP-60 and another hereditary condition with spastic paraplegia, named spastic paraplegia 13 (SPG13) and also due to mutation of the *hsp60* gene, present signs and symptoms resembling multiple sclerosis (MS). This should prompt clinicians and pathologists to think about the involvement of Hsp60 also in MS and stimulate research in this direction. This would be an example of the advantages of acquiring knowledge about chaperonopathies and, thus, improve differential diagnosis and patient management while providing a basis for research to elucidate the pathogenesis of diseases such as MS still not fully understood.

We consider that a diagnostic exercise does not conclude until the etiology of the syndrome under examination has been elucidated, and the pathogenesis, i.e., the mechanism by which the causal factor (the etiology) causes lesions and disease, has been understood. In our view, diagnosis is not just the labeling of a patient with the name of a known syndrome but it includes also the elucidation of etiology and pathogenesis. *Here, lies the importance of learning about chaperonopathies.* They are the etiologic-pathogenic factors in a variety of diseases as already established, and seem to be also implicated in many others. The latter require the special attention of, and diagnostic efforts by, clinicians and pathologists so the role of chaperonopathies in their pathogenesis can be confirmed or ruled out. One of the objectives of this book is to provide information to clinicians and pathologists so they can identify prospective chaperonopathies and proceed to study them to confirm or rule out their initial suspicion.

On the basis of current knowledge, a rule of thumb would be to think of abnormalities of: sHsp in peripheral nervous system syndromes and myopathies; Hsp60, Hsp70 and Hsp90 in cancers; Hsp60 in chronic inflammatory syndromes of the intestinal and respiratory tracts; and Hsp60 in autoimmune conditions as well as in age-associated generalized inflammatory syndromes. Also, a useful generalization would be to think of chaperonopathy when there are abnormalities in the formation of multimolecular assemblies such as microtubules, microfibrils, and muscle and collagen fibers, and when there is failure of interaction and function in multistep processes, such as that involving the respiratory chain in mitochondria. In addition, the presence of abnormal protein deposits, i.e., protein precipitates inside cells or in the intercellular space, is suggestive of chaperonopathy.

1.5 Diagnosis

It is unlikely that chaperonopathy will figure in the mind of most physicians when examining a patient because this nosological entity is not usually mentioned in current medical curricula or textbooks. Today's physicians are unprepared to diagnose and manage chaperonopathies. It is hoped that these deficiencies will soon disappear. This book is meant as a contribution to that end. Nowadays, the usual scenario is that of a patient with signs and symptoms of "well known" diseases, e.g., ulcerative colitis, Crohn's disease, chronic obstructive pulmonary disease (COPD), atherosclerosis, dilated cardiomyopathy, Charcot-Marie-Tooth disease, etc. who is examined according to current knowledge. The diagnostic exercise concludes with putting a name to the patient's condition without clarifying its etiology or pathogenic factors. Or if etiologic and pathogenic considerations are included in the final assessment of the patient, it is highly unlikely that the role of chaperones is considered. The physician will have to be trained to think that a chaperonopathy might be all or part of the cause of the patient's sufferings. But if chaperonopathies do figure in the list of conditions for differential diagnosis, then the pertinent course of action will soon become clear to achieve accurate diagnosis. A chaperonopathy will thus be diagnosed correctly or ruled out. This strategy will improve patient's management.

If a genetic chaperonopathy is suspected, analysis of the patient's DNA and that of relatives is indicated to detect mutations, for instance, in the chaperone gene suspected to be causing disease. When the signs and symptoms suggest acquired chaperonopathy, the suspected chaperone must be assessed in the most probable target tissues, and in biological fluids, to determine levels and other features. For instance, in tissue biopsies, the levels of the suspected chaperone(s) have to be determined, as well its location inside and outside the cell, including its association or colocalization with other molecules in pertinent cells types, such as macrophages and neutrophils in airways mucosa in COPD, just to mention one condition among the many in which chaperones are likely to play a pathogenic role.

1.6 Chaperones as Clinically Useful Biomarkers

Quantitative abnormalities of chaperones may be primary, namely they are the pathogenic factor, or secondary, namely they are the manifestation of another disease. The latter are signs of a disease that is not a primary chaperonopathy. In either case, assessment of chaperone levels in cells, tissues, and body fluids has proven to be a promising way to monitor disease progression and response to treatment. It is, therefore, necessary to include in the diagnostic work-up of many clinical conditions an evaluation of chaperones, or at least an evaluation of the chaperone most likely to be involved in the disorder under scrutiny. An example of such conditions is ulcerative colitis, in which it has been observed that the levels of Hsp60 in the affected colon mucosa are increased during relapse but decrease during remission and in response to treatment with improvement of the patient's status (see Sect. 7.4.2). Similarly, Hsp60 was found to decrease in airways mucosa as carcinogenesis progresses from initial hyperplasia and dysplasia toward fully developed carcinoma. Hence, it is justified to perform sequential evaluation of Hsp60 in cases such as those two examples mentioned in the preceding lines in biopsies of the pertinent mucosae, colon and airways, respectively.

1.7 Detection and Study of Chaperonopathies: What to Do?

The approach to studying chaperonopathies for diagnostic and clinical investigation purposes ought to include:

(1) Clinical description of symptoms to determine which organ or system is compromised and arrive at a preliminary clinical diagnosis of syndrome or disease; and (2) Clinical laboratory work-up as required by the preliminary clinical diagnosis.

If participation of chaperones in the etiology-pathogenesis of the patient's condition is suspected, one or more of the following steps are indicated:

1. Measurement of the suspected chaperone in the blood and, in specific situations, in other biological fluids such as cerebrospinal fluid, saliva, urine, sweat, tears, crevicular fluid, and genital secretion. Determination of the chaperone in the blood may require measurement of soluble forms in serum as well as in extracellular vesicles, e.g., exosomes, in plasma or serum.
2. Determination of the suspected chaperone in the tissue or tissues affected by resorting to biopsy (surgical, aspiration, punch, brush, etc. as required by the circumstances) and by applying immunocyto- and immunohistochemistry, immunofluorescence (including double immunofluorescence and other methods to establish if the chaperone is bound to, or colocalizes with, other molecules), and other techniques (e.g., immuno-electron microscopy) if deemed necessary. In immunocyto- and immunohistochemical analyses the levels and distribution of the chaperone has to be established in the affected cell/tissue and compared with non-affected counterparts. The cell types in which the chaperone is increased or decreased have to be identified. Localization of the chaperone in the cytosol, organelles (including nucleus), vesicles (intra- and extracellular), membrane, or extracellular space has to be established.
3. Determination of the suspected chaperone in lysates of the affected tissue, using Western blotting, immunoprecipitation, and proteomics (including two-dimensional (2-D) electrophoresis and mass spectrometry, for instance) to determine levels, electrophoretic characteristics in comparison with the normal tissue counterpart, and presence of post-translational modifications.

It is pertinent to bear in mind that quantitative assessment of chaperone levels in tissue lysates (and in cells and tissues as mentioned in the preceding point 2) may require not only measurement of the protein chaperone but also that of its corresponding mRNA, using for example quantitative RT-PCR (real-time polymerase chain reaction). The latter measurement provides information on the degree of expression of the chaperone gene, normal, elevated, or decreased; this information helps in interpreting the mechanism of the quantitative change observed at the protein level since it reveals if the gene is producing the chaperone at normal or abnormal levels.

In the study of the interaction of chaperones with other molecules in the cell various techniques can be used. One of them is the Chromatin Immunoprecipitation (ChIP) assay that allows, for example, determining if a chaperone interacts with DNA or with a protein bound to it.

4. Measurement of anti-chaperone antibodies in serum. It may be necessary to localize where the anti-chaperone antibody reacts, i.e., with what tissue or structure (e.g., vascular endothelium, myocardium, renal glomerulus, colon mucosa, airways mucosa) and whether the chaperone is part of the antigen–antibody complex. It may also be indicated to search for antibodies against a

foreign chaperone (e.g., Hsp60) from pathogens, such as *Chlamydia tracho-matis* and *Chlamydia pneumoniae*. Detection of such antibodies at levels definitely above those found in healthy individuals should direct the attention of the physician to chronic infections, relatively silent, that can initiate and maintain an autoimmune condition involving a chaperone (e.g., Hsp60), prolongation of which may cause serious and widespread lesions and, ultimately, disease.

5. Other pertinent determinations such as cytokines and products of immune-inflammatory system activation, and inflammatory parameters, using for instance flow cytometry. These determinations are particularly called for when a process of chronic, generalized inflammation is present, such as that observed in aged individuals, and chronic inflammatory syndromes of airways, digestive, and genito-urinary tracts. If the suspected chaperone is thought to play a role in activating the immune system and, thus, in pathogenesis, a pertinent therapeutic strategy should include blocking the chaperone (negative chaperonotherapy) to stop or limit its stimulation of the immune-inflammation response.

6. If a genetic chaperonopathy is suspected, genetic analysis should be carried out with the patient's DNA to search for gene mutations and polymorphisms along with the study of the patient's relatives to assess inheritance modality.

1.8 Chaperonotherapy

Chaperones can, in principle, be used in therapy, for instance, to replace a deficient chaperone. The chaperone protein or the gene can be utilized in replacement therapy in chaperonopathies by deficiency. Also, endogenous chaperone genes can be induced to produce chaperones using agents that preferentially induce those genes, such as heat shock and geranylgeranylacetone. On the contrary, in the chaperonopathies by mistake or collaborationism, when it is necessary to block or eliminate a chaperone which is promoting disease, anti-chaperone agents are indicated (negative chaperonotherapy). This topic will not be discussed in detail in this book; more information can be found in the publications listed in the pertinent section of Further Reading.

1.9 A Note on Bibliography

Suggested articles for Further Reading are listed at the end of each chapter. This list and the cited sources, including databases on line, throughout the book constitute a minimal but useful repository of information pertinent to chaperones and, most of all to chaperonopathies. Hsp60 is discussed often in the book because of the authors' direct experience with this chaperone.

1.10 The Future Near and Far

The study of chaperonopathies will surely expand in the near future. This expansion is likely to include elucidation of whether or not many candidate chaperonopathies, some presented in this book and others that will certainly be encountered by clinicians, pathologists, and geneticists, are genuine and can indeed be classified as such. In parallel, one may envision many studies aiming at clarifying the molecular mechanisms implicated in the development of the cell and tissue lesions observed in chaperonopathies and at unveiling the role played in these mechanisms by defective chaperones. For this purpose, model systems will have to be developed or standardized, using prokaryotes or simple eukaryotes, to circumvent the extreme complexity of the human chaperoning system, including the organellar subsytems and the multiple chaperoning teams and networks. Among the prokaryotes, archaeal organisms are favorites because their chaperones are more similar to those of humans than the bacterial counterparts.

Based on the findings from this research aiming at elucidating the details of the molecular mechanisms causing lesions and disease in chaperonopathies, one would expect that novel diagnostic and therapeutic means for these disorders will be invented in the long term. Development of these novel means to identify and treat chaperonopathies will necessarily require information on several aspects of the physiology of chaperones in health and disease. One of these aspects is the mechanism by which intracellular chaperones gain the intercellular space and then the circulation to reach destinations near to, or far away from, their cells of origin. To illustrate how elucidation of this mechanism might be achieved, the last chapter describes recent studies on secretion of Hsp60 by tumor cells. Along these lines, questions that will surely intrigue scientists and practioners alike are: Where do chaperones go after exiting their cells of origin? And, what do they do at their destinations? No doubt these and similar questions will inspire basic and clinical research in the short term.

Further Reading

Sections 1.1–1.5

Stress and Anti-Stress Mechanisms, Including Molecular Chaperones and Their Associated Pathologies
Encyclopedia of Stress (Second Edition). Editor-in-Chief: George Fink. Associate Editors: Bruce McEwen, E. Ronald de Kloet, Robert Rubin, George Chrousos, Andrew Steptoe, Noel Rose, Ian Craig, Giora Feuerstein. Copyright © 2007 Elsevier Inc. All rights reserved. ISBN: 978-0-12-373947-6. Elsevier Saint Louis, MO, USA; Oxford, UK; Singapore; Australia; http://www.elsevierdirect.com/brochures/stress/index.html

History

Haak J, Kregel KC (2008) 1962–2007: a cell stress odyssey. Novartis Found Symp 291:3–15; discussion 15–22, 137–140

CHAPERONES IN THE MAINTENANCE OF PROTEIN HOMEOSTASIS

Protein Folding, Protection from Aggregation, Dissolution of Protein Aggregates, Refolding, Translocation, Degradation, Namely, Chaperones in Protein Quality Control

Calamini B, Morimoto RI (2012) Protein homeostasis as a therapeutic target for diseases of protein conformation. Curr Top Med Chem 12:2623–2640

Clark AR, Lubsen NH, Slingsby C (2012) sHSP in the eye lens: Crystallin mutations, cataract and proteostasis. Int J Biochem Cell Biol 44:1687–1697

Easton DP, Kaneko Y, Subjeck JR (2000) The hsp110 and Grp170 stress proteins: newly recognized relatives of the Hsp70s. Cell Stress Chaperones 5:276–290

Hartl FU, Bracher A, Hayer-Hartl M (2011) Molecular chaperones in protein folding and proteostasis. Nature 475:324–332

Horwich AL, Fenton WA, Chapman E, Farr GW (2007) Two families of chaperonin: physiology and mechanism. Annu Rev Cell Dev Biol 23:115–145

Structure of Chaperonins: Ring-Shaped Oligomers

Fan M, Rao T, Zacco E, Ahmed MT, Shukla A, Ojha A, Freeke J, Robinson CV, Benesch JL, Lund PA (2012) The unusual mycobacterial chaperonins: evidence for in vivo oligomerization and specialization of function. Mol Microbiol 85:934–944

Chaperones and Autophagy

Arias E, Cuervo AM (2011) Chaperone-mediated autophagy in protein quality control. Curr Opin Cell Biol 23:184–189

Kaushik S, Cuervo AM (2012) Chaperones in autophagy. Pharmacol Res 66:484–493

Chaperones in Prevention of Aggregation of Defective Proteins

Athanasiou D, Kosmaoglou M, Kanuga N, Novoselov SS, Paton AW, Paton JC, Chapple JP, Cheetham ME (2012) BiP prevents rod opsin aggregation. Mol Biol Cell 23:3522–3531

Chaperones in Disaggregation of Protein Aggregates

Rosenzweig R, Moradi S, Zarrine-Afsar A, Glover JR, Kay LE (2013) Unraveling the mechanism of protein disaggregation through a ClpB-DnaK interaction. Science [Epub ahead of print] PMID: 23393091 [PubMed—as supplied by publisher]

Sharma SK, Christen P, Goloubinoff P (2009) Disaggregating chaperones: an unfolding story. Curr Protein Pept Sci 10:432–446

Winkler J, Tyedmers J, Bukau B, Mogk A (2012) Hsp70 targets Hsp100 chaperones to substrates for protein disaggregation and prion fragmentation. J Cell Biol 198:387–404

Zolkiewski M, Zhang T, Nagy M (2012) Aggregate reactivation mediated by the Hsp100 chaperones. Arch Biochem Biophys 520:1–6

Mitochondrial Chaperones

Tatsuta T (2009) Protein quality control in mitochondria. J Biochem 146:455–461
Voos W (2013) Chaperone-protease networks in mitochondrial protein homeostasis. Biochim
    Biophys Acta 1833:388–399. doi:10.1016/j.bbamcr.2012.06.005

Endoplasmic Reticulum (ER) Chaperones and ER-Associated Degradation (ERAD)

Benyair R, Ron E, Lederkremer GZ (2011) Protein quality control, retention, and degradation at
    the endoplasmic reticulum. Int Rev Cell Mol Biol 292:197–280
Ni M, Lee AS (2007) ER chaperones in mammalian development and human diseases. FEBS Lett
    581:3641–3651
Suntharalingam A, Abisambra JF, O'Leary JC 3rd, Koren J 3rd, Zhang B, Joe MK, Blair LJ, Hill
    SE, Jinwal UK, Cockman M, Duerfeldt AS, Tomarev S, Blagg BS, Lieberman RL, Dickey
    CA (2012) Grp94 triage of mutant myocilin through ERAD subverts a more efficient
    autophagic clearance mechanism. J Biol Chem 287:40661–40669

Dedicated Chaperones

Szolajska E, Chroboczek J (2011) Faithful chaperones. Cell Mol Life Sci 68:3307–3322

Histone Chaperones

Burgess RJ, Zhang Z (2013) Histone chaperones in nucleosome assembly and human disease. Nat
    Struct Mol Biol 2013 Jan; 20(1):14–22. doi:10.1038/nsmb.2461

Extracellular Chaperones

De Maio A (2011) Extracellular heat shock proteins, cellular export vesicles, and the stress
    observation system: a form of communication during injury, infection, and cell damage. It is
    never known how far a controversial finding will go! Dedicated to Ferruccio Ritossa. Cell
    Stress Chaperones 16:235–249
Henderson B, Pockley AG (2010) Molecular chaperones and protein-folding catalysts as
    intercellular signaling regulators in immunity and inflammation. J Leukoc Biol 88:445–462

Sections 1.6–1.7

CHAPERONES AS CLINICALY USEFUL BIOMARKERS

Cappello F, Conway de Macario E, Zummo G, Macario AJL (2011) Immunohistochemistry of
    human Hsp60 in health and disease: from autoimmunity to cancer. Methods Mol Biol
    787:245–254

The Value of Morphological-Topographic Data Obtained by Immunohistochemistry

Pellicciari C, Malatesta M (2011) Identifying pathological biomarkers: histochemistry still ranks
    high in the omics era. Eur J Histochem 55(4):e42; doi: 10.4081/ejh.2011.e42

Colon Cancer

Cappello F, David S, Peri G, Farina F, Conway de Macario E, Macario AJL, Zummo G (2011) The human chaperonin Hsp60: molecular anatomy, role in carcinogenesis and potential for diagnosis and treatment of colorectal cancer. Front Biosci S3:341–351

Cappello F, David S, Rappa F, Bucchieri F, Marasa L, Bartolotta TE, Farina F, Zummo G (2005) The expression of Hsp60 and Hsp10 in large bowel carcinomas with lymph node metastase. BMC Cancer 28(5):139

Dundas SR, Lawrie LC, Rooney PH, Murray GI (2005) Mortalin is over-expressed by colorectal adenocarcinomas and correlates with poor survival. J Pathol 205:74–81

Hamelin C, Cornut E, Poirier F, Pons S, Beaulieu C, Charrier JP, Haïdous H, Cotte E, Lambert C, Piard F, Ataman-Önal Y, Choquet-Kastylevsky G (2011) Identification and verification of HSP60 as a potential serum marker for colorectal cancer. FEBS J 278:4845–4859. doi:10.1111/j.1742-4658.2011.08385.x

Lung Cancer

Cappello F, Di Stefano A, D'Anna SE, Donner CF, Zummo G (2005) Immunopositivity of heat shock protein 60 as a biomarker of bronchial carcinogenesis. Lancet Oncol 6:816

Cappello F, Di Stefano A, David S, Rappa F, Anzalone R, La Rocca G, D'Anna SE, Magno F, Donner CF, Balbi B, Zummo G (2006) Hsp60 and Hsp10 down-regulation predicts bronchial epithelial carcinogenesis in smokers with chronic obstructive pulmonary disease. Cancer 107:2417–2424

Xu X, Wang W, Shao W, Yin W, Chen H, Qiu Y, Mo M, Zhao J, Deng Q, He J (2011) Heat shock protein-60 expression was significantly correlated with the prognosis of lung adenocarcinoma. J Surg Oncol 104:598–603. doi:10.1002/jso.21992

Prostate Cancer

Cappello F, Rappa F, David S, Anzalone R, Zummo G (2003) Immunohistochemical evaluation of PCNA, p53, HSP60, Hsp10 and MUC-2 presence and expression in prostate carcinogenesis. Anticancer Res 23:1325–1331

Glaessgen A, Jonmarker S, Lindberg A, Nilsson B, Lewensohn R, Ekman P, Valdman A, Egevad L (2008) Heat shock proteins 27, 60 and 70 as prognostic markers of prostate cancer. APMIS 116:888–895

Cardiovascular Disease

Novo G, Cappello F, Rizzo M, Fazio G, Zambuto S, Tortorici E, Marino Gammazza A, Corrao S, Zummo G, Conway de Macario E, Macario AJL, Assennato P, Novo S, Li Volti G (2011) Hsp60 and Heme Oxygenase-1 (Hsp32) in acute myocardial infarction. Trans Res 157:285–292

Rizzo M, Macario AJL, Conway de Macario E, Gouni-Berthold I, Berthold H, Rini GB, Zummo G, Cappello F (2011) Heat-shock protein 60 and risk for cardiovascular disease. Curr Pharmaceutical Des 17:3662–3668

Inflammatory Bowel Disease

Rodolico V, Tomasello G, Zerilli M, Martorana A, Pitruzzella A, Gammazza AM, David S, Zummo G, Damiani P, Accomando S, Conway de Macario E, Macario AJL, Cappello F (2010) Hsp60 and Hsp10 increase in colon mucosa of Crohn's disease and ulcerative colitis. Cell Stress Chaperones 15:877–884

Seigneuric R, Mjahed H, Gobbo J, Joly A-L, Berthenet K, Shirley S, Garrido C (2011) Heat shock proteins as danger signals for cancer detection. Front Oncol; 1:37, doi: 10.3389/fonc.2011.00037

Sections 1.8 and 1.10

Chaperonotherapy and the Future

Abisambra JF, Jinwal UK, Jones JR, Blair LJ, Koren J 3rd, Dickey CA (2011) Exploiting the diversity of the heat-shock protein family for primary and secondary tauopathy therapeutics. Curr Neuropharmacol 9:623–631

Almeida MB, do Nascimento JL, Herculano AM, Crespo-López ME (2011) Molecular chaperones: toward new therapeutic tools. Biomed Pharmacother 65:239–243

Cappello F, Czarnecka AM, Rocca GL, Stefano AD, Zummo G, Macario AJL (2007) Hsp60 and Hsp10 as antitumor molecular agents. Cancer Biol Ther 6:487–489

Cappello F, Angileri F, Conway de Macario E, Macario AJL (2013) Chaperonopathies and chaperonotherapy. Hsp60 as therapeutic target in cancer: potential benefits and risks. Curr Pharm Des 19:452–457

Deocaris CC, Kaul SC, Wadhwa R (2009) The versatile stress protein mortalin as a chaperone therapeutic agent. Protein Pept Lett 16:517–529

Goloudina AR, Demidov ON, Garrido C (2012) Inhibition of HSP70: a challenging anti-cancer strategy. Cancer Lett 325:117–124

Kalmar B, Greensmith L (2009) Induction of heat shock proteins for protection against oxidative stress. Adv Drug Deliv Rev 61:310–318

Lee J, Tan CY, Lee SK, Kim YH, Lee KY (2009) Controlled delivery of heat shock protein using an injectable microsphere/hydrogel combination system for the treatment of myocardial infarction. J Control Release 137:196–202

Pace A, Barone G, Lauria A, Martorana A, Palumbo Piccionello A, Pierro P, Terenzi T, Almerico AM, Buscemi S, Campanella C, Angileri F, Carini F, Zummo G, Conway de Macario E, Cappello F, Macario AJL (2013) Hsp60, a novel target for antitumor therapy: Structure-function features and prospective drugs design. Curr Pharm Des 19:2757–2764

Raju M, Santhoshkumar P (2012) Sharma KK (2012) αA-Crystallin-derived mini-chaperone modulates stability and function of cataract causing αAG98R-Crystallin. PLoS ONE 7(9):e44077

Tan MH, Smith AJ, Pawlyk B, Xu X, Liu X, Bainbridge JB, Basche M, McIntosh J, Tran HV, Nathwani A, Li T, Ali RR (2009) Gene therapy for retinitis pigmentosa and Leber congenital amaurosis caused by defects in AIPL1: effective rescue of mouse models of partial and complete Aipl1 deficiency using AAV2/2 and AAV2/8 vectors. Hum Mol Genet, 18:2099–2114. Erratum in. Hum Mol Genet 19(4):735

Vizcaychipi MP, Xu L, Barreto GE, Ma D, Maze M, Giffard RG (2011) Heat shock protein 72 overexpression prevents early postoperative memory decline after orthopedic surgery under general anesthesia in mice. Anesthesiology 114:891–900

White RE, Ouyang Y-B, Giffard R (2012) Hsp75/mortalin and protection from ischemic brain injury. In: Kaul SC, Wadhwa R (eds) Mortalin Biology: life, stress and death, Springer, Berlin, pp 179–190

Pharmacological Chaperones

Guce AI, Clark NE, Rogich JJ, Garman SC (2011) The molecular basis of pharmacological chaperoning in human α-galactosidase. Chem Biol 18:1521–1526

Chaperone Inhibitors

da Silva VC, Ramos CH (2012) The network interaction of the human cytosolic 90kDa heat shock protein Hsp90: A target for cancer therapeutics. J Proteomics 75:2790–2802

Gomez-Monterrey I, Sala M, Musella S, Campiglia P (2012) Heat shock protein 90 inhibitors as therapeutic agents. Recent Pat Anticancer Drug Discov 7:313–336

Lee CH, Hong HM, Chang YY, Chang WW (2012) Inhibition of heat shock protein (Hsp) 27 potentiates the suppressive effect of Hsp90 inhibitors in targeting breast cancer stem-like cells. Biochimie 94:1382–1389

Archaeal and Eukaryal Chaperones are Evolutionary Related and Similar. Hence, Archaea have Potential as Experimental Models Simpler and, Thus, More Amenable to Dissection and Analysis than Eukaryotic Counterparts to Study Human Chaperones and the Effect of Mutations on Them

Conway de Macario E, Maeder DL, Macario AJL (2003) Breaking the mould: Archaea with all four chaperoning systems. Biochem Biophys Res Commun 301:811–812

Large AT, Lund PA (2009) Archaeal chaperonins. Front Biosci 14:1304–1324. To see Table of Contents, Abstract, Figures and Tables go to: http://www.bioscience.org/2009/v14/af/3310/fulltext.htm

Luo H, Laksanalamai P, Robb FT (2009) An exceptionally stable group II chaperonin from the hyperthermophile *Pyrococcus furiosus*. Arch Biochem Biophys 486:12–18

Macario AJL, Conway de Macario E (2001) The molecular chaperone system and other anti-stress mechanisms in archaea. Front Biosci 6:d262–283. To see Table of Contents, Abstract and Figures go to: http://www.bioscience.org/2001/v6/d/macario/fulltext.htm

Maeder DL, Macario AJL, Conway de Macario E (2005) Novel chaperonins in a prokaryote. J Mol Evol 60:409–416

Chapter 2
Chaperones: General Characteristics and Classifications

Abstract This chapter presents the classification of chaperones, their molecular properties among which that of forming functional complexes involving various molecules, and their distribution inside and outside the cell. The chaperone genes in the human genome are listed and briefly described, focusing on the small heat-shock proteins (sHsp), Hsp60, Hsp70, and Hsp90, and mentioning all others known. The chapter also introduces the concept of chaperoning system, i.e., the physiological system of an organism which is composed of all its chaperones, co-chaperones, and chaperone co-factors.

Keywords Chaperone genes, human genome · sHsp · Hsp40/DnaJ · Hsp60 · Hsp70 · Hsp90 · Chaperoning teams · Chaperoning networks · Chaperoning system · Ubiquitin-proteasome system (UPS) · Apoptosis

2.1 Classification of Chaperones

Hsp-chaperones are classified in various ways. For example, according to molecular weight they fall into several groups, Table 2.1. In some of these groups are classified well known families of phylogenetically related proteins, such as the Hsp90, Hsp70, Hsp40, and sHsp-crystallin families. Other members of these groups are not members of those families although they have molecular weights within the stipulated ranges. Chaperonopathies involve chaperones pertaining to all these groups.

The classification of chaperones according to their apparent molecular weight (MW) as determined, for instance, by gel electrophoresis is quite useful. If one knows the MW of the chaperone under scrutiny and wants to check if it is present in a sample (serum or another biologic fluid, secretion, or tissue extract), a simple gel electrophoresis with protein staining might give the first piece of information desired. Subsequent steps applying complementary methods such as Western blotting (using specific anti-chaperone antibodies; see for example Fig. 9.1), 2D electrophoresis and mass spectrometry will definitely clarify the situation. This is very encouraging because the need to determine presence, levels, and distribution of

A. J. L. Macario et al., *The Chaperonopathies*, SpringerBriefs in Biochemistry
and Molecular Biology, DOI: 10.1007/978-94-007-4667-1_2, © The Author(s) 2013

Table 2.1 Subpopulations of Hsp-chaperones classified according to apparent molecular weight

MW (kDa) range	Classical family in this MW range	Other members in this range implicated in the causation of disease (chaperonopathies)[a]
200 or higher	None	Sacsin
100–199	Hsp100–110	
81–99	Hsp90	Paraplegin [SPG7]
65–80	Hsp70/DnaK	Spastin [SPG4]; LARP7
55–64	Hsp60 (chaperonins groups I and II, e.g., Cpn60 and CCT, respectively)	Myocilin; protein disulphide-isomerases (PDI)
35–54	Hsp40/DnaJ	AIP; AIPL1; torsin A; clusterin; DNAJC19 (TIM14)
34 or less	sHsp (crystallins)	Hsp10 (Cpn10); Alpha hemoglobin-stabilizing protein (AHSP); cyclophilin type peptidyl-prolyl cis–trans isomerase (PPI); alpha-synuclein; HSPB11

[a] Others are named in the tables, for example Table 4.3, in which the pertinent chaperonopathies are listed

Source Macario AJL (1995) Heat-shock proteins and molecular chaperones: Implications for pathogenesis, diagnostics, and therapeutics. Intl J Clin Lab Res 25:59–70; and Macario AJL, Conway de Macario E (2009) The chaperoning system: Physiology and pathology. Exp Med Rev Vol. 2–3: Years 2008/09, pp. 9–21 (http://www.unipa.it/giovanni.zummo)

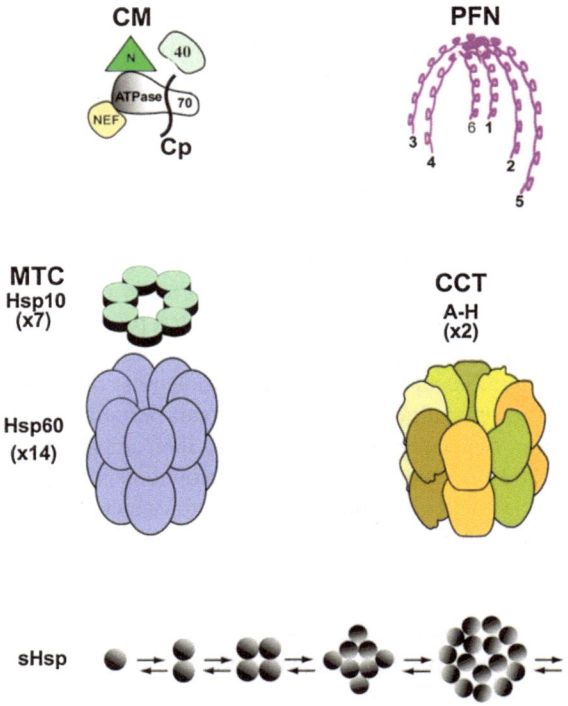

chaperones in clinical and other biological samples in research is quite common. It has to be borne in mind, though, that the electrophoretic characteristics of a chaperone may vary due to structural changes caused by post-translational modifications or mutations. In the case of deletions, the length of the protein may be considerably altered. So, the results have to be examined very critically, particularly when a chaperonopathy is suspected and the chaperone molecule might by abnormal.

2.2 Chaperones can Form Functional Complexes that Work as Multimolecular Machines

Chaperones usually do not work alone but form teams, sometimes called chaperoning machines. Teams can be formed by varios members of the same family, i.e., oligomers, in which all members are identical, homo-oligomers, or not,

◀ **Fig. 2.1 Examples of chaperoning teams. CM.** The chaperoning machine (*CM*) is a team of essentially three proteins, Hsp70 (70), Hsp40 (40), and nucleotide (N; e.g., adenosine triphosphate, abbreviated *ATP*) exchange factor (NEF). Hsp70 binds a client (e.g., unfolded or misfolded) polypeptide (*Cp*) and, in collaboration with the other members of the team, assists the polypeptide to fold correctly to achieve its native conformation. This chaperoning process involves ATP hydrolysis mediated by the ATPase domain of Hsp70 and stimulated by Hsp40, and exchange of adenosine diphosphate (*ADP*) for ATP, an exchange that is induced by the nucleotide exchange factor (e.g., BAG-1, which means Bcl-2-associated athanogene 1, where Bcl-2 stands for B cell lymphoma 2). See Fig. 2.2. **PFN.** Prefoldin (*PFN*) is composed of distinct subunits (1 through 6 in humans) arranged in a medusa type of structure. **MTC.** The mitochondrial chaperonin (*MTC*) is a barrel shaped complex of two multimeric assemblages. The larger assemblage is formed of two superimposed rings, each with seven identical subunits (Hsp60 also named Cpn60, where Cpn stands for chaperonin), delimiting a central cavity, the folding chamber. The smaller assemblage is also ring shaped, being formed by seven identical subunits (Hsp10, also named Cpn10) that are considerably smaller than those around the folding chamber. The *smaller ring* serves as a lid to close the folding chamber while the polypeptide folding process is taking place inside. **CCT.** The chaperonin-containing TCP-1 poypeptide (*CCT*), where TCP-1 means tailless complex polypeptide 1, is similar in overall structure to the mitochondrial chaperonin since it is also formed of two superimposed multisubunit rings and it has a central cavity. However, the CCT rings are formed of eight distinct subunits (A through H, also named alpha through theta) rather than seven identical subunits, and there is no third ring or lid. **sHsp.** The small heat-shock proteins (sHsp) are chaperones, such as the alpha-crystallins, that are usually 34 kDa or less in molecular weight (see Table 2.1) and are normally present as monomers, or multimers of varios degrees of complexity, depending on which sHsp, cell type, and conditions (e.g., stress vs. basal, constitutive) is considered. *Source* Macario AJL, Conway de Macario E (2005) Sick chaperones, cellular stress and disease. New Eng J Med 353:1489–1501; Macario AJL, Conway de Macario E (2001) The molecular chaperone system and other anti-stress mechanisms in archaea. Front Biosci 6: d262–283. To see Table of Contents, Abstract, Figures, and Tables go to http://www.bioscience.org/2001/v6/d/macario/fulltext.htm; Macario AJL, Malz M, Conway de Macario E (2004) Evolution of assisted protein folding: The distribution of the main chaperoning systems within the phylogenetic domain Archaea. Front Biosci 9: 1318–1332. To see Table of Contents and Abstract go to http://www.bioscience.org/2004/v9/af/1328/fulltext.htm; and Macario, AJL, Conway de Macario E (2007) Molecular chaperones: Multiple functions, pathologies, and potential applications. Front Biosci 12: 2588–2600. To see Table of Contents, Abstract and Figures go to http://www.bioscience.org/2007/v12/af/2257/fulltext.htm

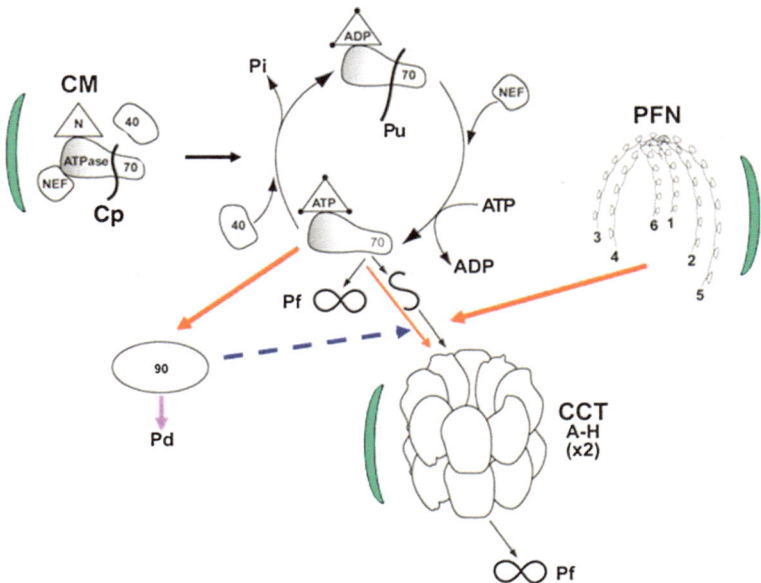

hetero-oligomers. An example of the former is the mitochondrial chaperonin Hsp60 complex, which is formed by seven identical subunits, Fig. 2.1, MTC, Hsp60. An example of hetero-oligomer is the chaperonin of Group II present in the cytosol, named chaperonin-containing TCP-1 polypeptide (CCT), Fig. 2.1.

Also shown in Fig. 2.1 are prefoldin (PFN), which comprises six different but very similar subunits in humans; Hsp10 (that builds heptamers; MTC, Hsp10), and other small heat-shock proteins, indicated as sHsp; and Hsp70 (ATPase 70).

Hsp70 forms a chaperoning machine (CM) by interacting with a co-chaperone, Hsp40/Dnaj (40) and with a co-factor, nucleotide exchange factor (NEF) and, thus, with participation of ATP (N) folds nascent polypeptides (client polypeptide, Cp), as illustrated in Fig. 2.2.

It is believed that the sHsp, or at least some of them, undergo dynamic assembly into mono- and poly-disperse oligomers (Fig. 2.1) with variable chaperoning ability. For example, the alpha-A crystallin (HspB4 in Table 2.3), the largest being composed of 30–40 identical subunits; this results in a highly dynamic quaternary structure with the subunits exchanging between the oligomers of various multi-plicities. The chaperoning ability seems to depend on the rate of oligomer disas-sembly. Another sHsp, Hsp27, may occur as multimers under basal conditions but the complexes disaggregate upon cellular stress and the free monomers become involved in counteracting the consequences of stress such as protein aggregation.

It becomes clear from these considerations and from the illustrations in Figs. 2.1, 2.2 that each single molecule of a chaperone must have different functional domains to carry out the various functions pertaining to team building and polypeptide folding. In addition, teams interact between them to build

◀ **Fig. 2.2 Examples of chaperoning teams forming a network.** The Hsp70–Hsp40-nucleotide-exchange factor (*NEF*) chaperone machine (*CM*) is a dynamic complex or team involved in protein folding that also interacts with other teams. *Top left* Hsp70 (70) binds the unfolded client polypeptide (*Cp*; represented as a *slightly ondulating line*), via its peptide-binding domain near the *middle* of the molecule, when ADP is bound to its ATPase domain and Hsp40 (40) is bound to its C-terminal domain. *Top middle* the client polypeptide, still unfolded (Pu) but already bound to Hsp70, enters the folding cycle. The nucleotide-exchange factor (e.g., BAG-1, or GrpE in prokaryotes, shown as NEF) promotes exchange of ADP for ATP. When Hsp70 is bound to ATP, its affinity for the folding polypeptide decreases, and the folded polypeptide (Pf; represented as ∞) is released. The nucleotide-exchange factor is replaced by Hsp40, ATP hydrolysis occurs with release of pyrophosphate (*Pi*), and a new cycle of peptide binding, folding and release begins. N, nucleotide, e.g., ATP or ADP. *The bottom half of the figure* shows other alternatives. For example, the partially folded polypeptide (represented as *S*) is handed over by the CM to the CCT complex, directly or with participation of Hsp90 (90; to the left) and/or prefoldin (PFN; to the right). The polypeptide is then folded inside the CCT-folding chamber and released. CM and Hsp90 can also be involved in a pathway leading to protein degradation (*Pd*) in the proteasome, for instance. While CM occurs in the three life Domains (i.e., Bacteria, Archaea, and Eukarya), CCT and PFN exist in eukaryotes (typically the cytosol) and archaea but not in bacteria. CM also exists in the eukaryotic cell compartments: nucleus, mitochondria, endoplasmic reticulum, and chloroplasts. *Red arrows* and blue discontinuous arrow indicate possible interactions between chaperoning teams to form a functional network. The *violet arrow* indicate the pathway to degradation by intracellular proteases, including the ubiquitin–proteasome system. The *green vertical arc* near each team indicate that the team members interact with each other and that the whole complex is dynamic, i.e., it changes conformation while performing its function. See also Sect. 7.2 for a brief discussion on the interaction of the chaperonin system with the ubiquitin–proteasome system (*UPS*) involved in protein degradation, and with autophagy, particularly chaperone-mediated autophagy (*CMA*), and with apoptosis (programmed cell death). *Source* Macario AJL, Conway de Macario E (2001) The molecular chaperone system and other anti-stress mechanisms in archaea. Front Biosci 6: d262–283. To see Table of Contents, Abstract and Figures go to http://www.bioscience.org/2001/v6/d/macario/fulltext.htm; Macario AJL, Malz M, Conway de Macario E (2004) Evolution of assisted protein folding: The distribution of the main chaperoning systems within the phylogenetic domain Archaea. Front Biosci 9: 1318–1332. To see Table of Contents and Abstract go to http://www.bioscience.org/2004/v9/af/1328/fulltext.htm; and Macario, AJL, Conway de Macario E (2007) Molecular chaperones: Multiple functions, pathologies, and potential applications. Front Biosci 12: 2588–2600. To see Table of Contents, Abstract, and Figures go to http://www.bioscience.org/2007/v12/af/2257/fulltext.htm

functional networks. Therefore, individual molecules must have domains dedicated to form networks. *It is then evident that there are many sites in the chaperone molecules in which a change due to mutation, or to post-translational modification, may abolish or alter one or more of their interactions and functions with the potential of causing a chaperonopathy.*

2.3 Distribution of Chaperones

Chaperones are found in all cells, cell compartments, tissues, and outside cells in the intercellular space, in circulation, and in secretions (Table 2.2). Consequently, it is to be expected that a chaperonopathy might have an impact on various locations, generating pleiotropic and variable pathological and clinical pictures.

Table 2.2 Residences of chaperones in eukaryotes

Location	Compartment
Intracellular	Nucleus; cytosol; mitochondria; endoplasmic reticulum (ER); lysosomes; vesicles; plasma membrane; chloroplasts (in plants)
Pericellular	Cell membrane on the outside
Extracellular	Intercellular space; blood (plasma, serum): soluble or in vesicles (e.g., exosomes); lymph; cerebrospinal fluid; intrasynovial space (joint cavity); secretions (e.g., saliva, urine)

Source Macario AJL, Conway de Macario E (2009) The chaperoning system: physiology and pathology. Exp Med Rev Vol. 2–3: Years 2008/09, pp. 9–21, (http://www.unipa.it/ giovanni.zummo); and Cappello F, Conway de Macario E, Marasa L, Zummo G, Macario AJL (2008) Hsp60: new locations, functions, and perspectives for cancer diagnosis and therapy. Cancer Biol Ther 7:801–809

It has become clear from recent studies that chaperones thought to be confined to a single cell compartment, e.g., the mitochondria, can also be found in other locations. It is of particular interest that several chaperones are now known to be on the plasma-cell membrane, exposed to the outside of the cell, especially in tumors. This knowledge opens new avenues for research and for developing diagnostic means centered on detecting tumors by targeting chaperones on their cells using antibodies. Even more exciting is the possibility of utilizing these antibodies to deliver anticancer drugs to the cancer cells, or to kill them immunologically (see also Chap. 9).

As a direct consequence of the re-assessment of the distribution of chaperones intracellularly and extracellularly and its alterations in disease, morphological methods that detect chaperones in situ have been re-evaluated and given considerable importance. Immunocytochemistry and immunohistochemistry with specific antibodies, and complementary morphological and quantitative techniques (e.g., confocal microscopy), are instrumental to estimate chaperone levels in cells and tissues and their changes with disease progression and response to treatment, for instance. These morphological methods are also very useful to map the distribution of chaperones in cells and tissues, and its variations in pathological situations such as the chaperonopathies. If electron microscopy is added, the immunogold technique with specific antibodies can provide crucial information on the chaperone location, for instance plasma-cell membrane, mitochondria, Golgi apparatus, or cytosol (see also Sect. 7.4 and Chap. 9).

2.4 How Many Chaperones in the Human Species?

Given that molecular chaperones can be affected by abnormalities, genetic or acquired, structural and/or functional, it is pertinent to ask how many chaperones are there in humans that might cause problems. How many chaperonopathies are possible? The answer to this question has become accessible since the development of strategies and methods for genome sequencing and analysis. We developed a protocol, chaperonomics, for identifying chaperone genes and, applying this protocol, we elucidated the complement of Hsp60/CCT and Hsp70 genes in the human genome. Likewise, others have identified other members of the chaperoning system in humans. It is likely that more members will be discovered as the understanding of the genome progresses and the boundaries of genes and the extent of regulatory regions are better defined. In any case, the information already available provides an excellent basis to study chaperones and their abnormalities or chaperonopathies. In addition, the diversity of chaperones stemming from the set of genes in the genome is most likely enlarged by, for instance, variations in the transcription initiation site, differential (alternative) splicing, and post-translational modifications. This diversity of mechanisms involved in the synthesis and production of chaperone molecules could be the basis of the diverse functions and locations that are constantly being unveiled for the chaperone proteins (see Table 2.2, and Sect. 2.5). These transcriptional and post-translational variations will, no doubt, be the focus of research in the immediate future because they have the potential of uncovering fundamental principles of biology in what regards molecular migration and diversification of function.

Unfortunately, many chaperones have been given various names, which can be very confusing and frustrating. To assist the reader, we have included in some of

Table 2.3 Human sHsp genes and proteins

Alpha-crystallin family	New nomenclature		Other names	Gene ID	aa
	Gene	Protein			
HspB1	*HSPB1*	HSPB1	CMT2F; HMN2B; HSP27; HSP28; HSP25; HS.76067; DKFZp586P1322	3315	205
HspB2	*HSPB2*	HSPB2	MKBP; HSP27; Hs.78846; LOH11CR1K; MGC133245	3316	182
HspB3	*HSPB3*	HSPB3	HSPL27	8988	150
HspB4	*HSPB4*	HSPB4	Crystallin alpha A; CRYAA, CRYA1	1409	173
HspB5	*HSPB5*	HSPB5	Crystallin alpha B, CRYAB; CRYA2	1410	175
HspB6i	*HSPB6*	HSPB6	HSP20; FLJ32389	126393	160
HspB7	*HSPB7*	HSPB7	cvHSP; FLJ32733; DKFZp779D0968	27129	170
HspB8	*HSPB8*	HSPB8	H11; HMN2; CMT2L; DHMN2; E2IG1; HMN2A; HSP22; CRYAC	26353	196
HspB9	*HSPB9*	HSPB9	FLJ27437	94086	159
HspB10	*HSPB10*[a]	HSPB10	ODF1; ODF; RT7; ODF2; ODFP; SODF; ODF27; ODFPG; ODFPGA; ODFPGB; MGC129928; MGC129929	4956	250
	HSPB11	HSPB11	HSP16.2; C1orf41; PP25; IFT25; HSPCO34	51668	144

Source Kappe G, Franck E, Verschuure P, Boelens WC, Leunissen JAM, de Jong WW (2003) The human genome encodes 10 alpha-crystallin–related small heat shock proteins: HspB1–10. Cell Stress Chaperones 8:53–61; Kampinga HH, Hageman J, Vos MJ, Kubota H, Tanguay RM, Bruford EA, Cheetham ME, Chen B, Hightower LE (2009) Guidelines for the nomenclature of the human heat shock proteins. Cell Stress Chaperones 14: 105–111; doi: 10.1007/s12192-008-0068-7; and Hu Z, Yang B, Lu W, Zhou W, Zeng L, Li T, Wang X (2008) HSPB2/MKBP, a novel and unique member of the small heat-shock protein family. J Neurosci Res 86:2125–2133

the Tables describing chaperones not only the preferred name but also some other names that are, or have been, used in data bases and publications.

2.4.1 Human Small Size (34 kDa or Less) Chaperones

Thus far 10 genes encoding small heat shock proteins (sHsp) with the crystallin domain have been identified in the human genome (Table 2.3). However, there are several other chaperones with a MW similar to those of the sHsp (e.g., HspB11 in Table 2.3, and many others some of which are listed in Table 2.1), which can be affected by pathologic changes and, thus, can be the basis of a chaperonopathy.

Table 2.4 Human *hsp60/CCT* genes

Name	Alternative names	Start	End	Str	Chr	Loc	Is	Ex	aa
CCT1	TCP1, CCTa, TCP-1 alpha	160,119,520	160,130,731	−	6	q25.3	2	12, 7	556, 401
CCT2	TCP1 beta	68,266,317	68,280,052	+	12	q15	1	14	535
CCT3	TCP1 gamma	154,545,617	154,572,307	−	1	q23.1	3	13, 13, 12	545, 544, 507
CCT4	TCPD, TCP-1 delta	61,950,076	61,969,146	−	2	p15	1	13	539
CCT5	TCP1E, TCP1 epsilon, KIAA0098	10,303,453	10,317,892	+	5	p15.2	1	11	541
CCT6A	TCP1 zeta, CCT6, Cctz, HTR3, TCP20, TCPZ, TTCP20	56,087,036	56,098,269	+	7	p11.2	2	14, 13	531, 486
CCT6B	Cctz2, TSA303, Tcp20	30,279,183	30,312,525	−	17	q12	1	14	530
CCT7	TCP1 eta	73,320,279	73,333,494	+	2	p13.2	2	12, 7	543, 339
CCT8	TCP1 theta	29,350,670	29,367,782	−	21	q21.3	1	15	548
CCT8L1	LOC155100	151,773,495	151,775,165	+	7	q36.1	1	1	557
CCT8L2	GROL, CESK1	15,451,770	15,453,440	−	22	q11.1	1	1	557
MKKS	BBS6	10,333,898	10,342,162	−	20	p12.2	2	4, 4	570, 570
BBS10	C12orf58, FLJ23560	75,263,727	75,266,269	−	12	q21.2	1	2	723
BBS12	C4orf24, FLJ35630, FLJ41559	123,882,498	123,884,627	+	4	q27	1	1	710
HSPD1	GROEL, HSP60, SPG13, CPN60, HuCHA60	198,060,018	198,071,817	−	2	q33.1	2	11, 11	573, 573
PIKFYVE	CFD, FAB1, PIP5 K, PIP5K3	209,182,591	209,190,094	+	2	q34	1	5	224

Key Name, official NCBI Entrez gene database name. For CCT1, the official Entrez name is TCP1 but we chose CCT1 for consistency with other subunit gene names; **Start** and **End**, start and end of coding region; **Str**, DNA strand with positive or negative signs indicating sequenced or complementary strand, respectively; **Chr**, chromosome; **Loc,** chromosomal location; **Is**, number of isoforms or mRNA variants; **Ex**, number of exons-multiple numbers indicate the number of exons in each isoform; **aa**, amino acids encoded; **(PIKFYVE)**, Fab1_TCP sequence domain of the PIKFYVE kinase, most similar to the apical domain of CCT3, in which features refer to the domain portion of the gene/protein

Source Brocchieri L, Conway de Macario E, Macario AJL (2007) Chaperonomics, a new tool to study ageing and associated diseases. Mechan Ageing Develop. 128:125–136; and Mukherjee K, Conway de Macario E, Macario AJL, Brocchieri L (2010) Chaperonin genes on the rise: new divergent classes and intense duplication in human and other vertebrate genomes. BMC Evolutionary Biology 2010, 10:64; doi:10.1186/1471-2148-10-64. http://www.biomedcentral.com/1471-2148/10/64

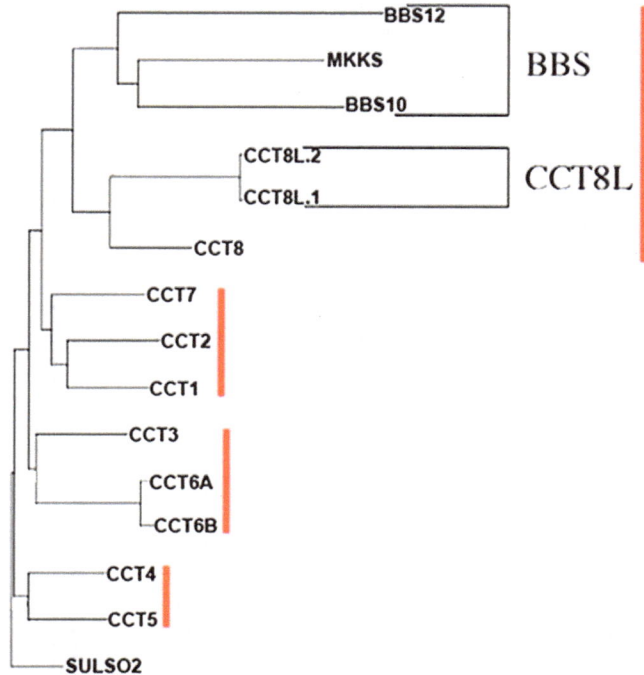

Fig. 2.3 The branching patterns of the canonical (CCT1–CCT8) subunits in the maximum likelihood tree indicate that distinct clades are formed by the subunits CCT4 and CCT5, by the subunits CCT3 and CCT6, and, with strong bootstrap support, by the subunits CCT1, CCT2 and CCT7. The data also indicate that the BBS clade and CCT8L share a common ancestor with CCT8 but are not necessarily derived from CCT8. The tree was rooted by the archaeal thermosome alpha subunit of *Sulfolobus solfataricus* (SULSO2). *Source* Mukherjee K, Conway de Macario E, Macario AJL, Brocchieri L (2010) Chaperonin genes on the rise: new divergent classes and intense duplication in human and other vertebrate genomes. BMC Evolutionary Biology 2010, 10:64; doi:10.1186/1471-2148-10-64. http://www.biomedcentral.com/1471-2148/10/64

2.4.2 Human Hsp60/CCT Genes and Proteins

The human genome harbors 16 *hsp60/CCT* genes, including the mitochondrial Hsp60 or Cpn60 (Table 2.4) and many related pseudogenes. The CCT genes form 4 (or 5, depending on the criteria applied) evolutionarily related groups (Fig. 2.3).

2.4.3 Human Hsp70 Genes and Proteins

The human genome harbors 17 *hsp70* genes (Table 2.5, parts 1 and 2) and many related pseudogenes. The genes form seven evolutionarily related groups (Fig. 2.4).

Table 2.5 *hsp70* genes in the human genome

Name	Location	Str	Start/End	nt	Ex	Is	aa
Part 1							
Hsp70kDa 6 (HSP70B')/HSPA6	1q23.3	+	159,761,073/ 159,763,001 N	1,929	1	1	643
Hsp70kDa 7 (HSP70B)/HSPA7	1q23.1	+	159,842,705/159,844,628	1,924	1	1	641
Hsp70kDa 4-like/HSPA4L	4q28.1	+	128,923,156/ 128,973,476 N	50,321	19	1	839
Hsp70kDa 9B/HSPA9B	5q31.2	–	137,919,628/137,938,906	19,279	17	1	679
Hsp70kDa 4/HSPA4	5q31.1	+	132,415,842/132,468,024	52,183	19	2; a	840
			132,415,842/132,468,024	52,183	5	b	148
Hsp70kDa 1-like/HSPA1L/HSP70-Hom	6p21.33	–	31,885,806/31,887,728	1,923	1	1	641
Hsp70kDa 1A/HSPA1A/HSP70-1	6p21.33	+	31,891,513/31,893,435	1,923	1	1	641
Hsp70 kDa 1B/HSPA1B/HSP70-2	6p21.32	+	31,903,707/31,905,629	1,923	1	1	641
Hsp70 kDa 5 (Grp78; BiP)/HSPA5	9q33.3	–	127,038,695/127,043,226	4532	8	1	654
Part 2							
Hsp70kDa 12A/HSPA12A	10q25.3	–	118,424,285/118,456,787	32,503	12	1	675
Hsp70kDa 14/HSPA14	10p13	+	14,920,408/14,953,608	33,201	14	2	509
			14,920,408/14,924,181	3,774	4		88
Hsp70kDa 8/HSPA8	11q24.1	–	122,433,655/122,437,242	3,588	8	2	646
			122,433,655/122,437,242	3,588	7		493
150kDa oxygen-regulated protein/ HYOU1	11q23.3	–	118,421,518/118,432,082	10,565	25	4	999
			118,421,518 / 118,432,082	10,565	25	[3]	999
			118,421,518/118,432,082	10,234	24		964
			<118,424,940/118,432,082	>7,143	16		687

(continued)

Table 2.5 (continued)

Name	Location	Str	Start/End	nt	Ex	Is	aa
Hsp105kD/HSPH1	13q12.3	–	30,609,458/30,633,719	24,262	18	2; α	858
			30,609,458/30,633,719	24,262	17	β	814
Hsp70kDa 2/HSPA2	14q23.3	+	64,077,321/64,079,237	1,917	1	1	639
Hsp70kD 12B/HSPA12B	20p13	+	3,667,322/3,680,810	13,489	12	1	686
Stress 70 protein chaperone/STCH	21q11.2	–	14,667,812/14,677,311	9,500	5	1	471

Parts 1 and 2. Human *hsp70* genes. *Key Name*, official NCBI Entrez gene database name; **Str**, DNA strand with positive or negative signs indicating sequenced or complementary strand, respectively; **Location**, chromosomal location as per UCSC genomic browser; **Start/End**, genomic positions at the 5′ and 3′ ends of the mRNA, including the 5′-UTR and 3′-UTR, except for HSPA7, for which the coding-region length is shown since there is no mRNA sequence available for this gene. The data are as per the UCSC genomic browser website; **nt**, gene length in nucleotides (base pairs); **Ex**, exons, excluding non-coding exons when present; **Is**, protein isoforms and mRNA variants, as pertinent. For the HYOU1 gene four mRNA variants and three protein isoforms (shown within brackets) have been determined; **aa**, amino acids encoded. HSPA7 was found to be transcribed under heat-shock in fibroblasts and it was concluded that the gene is functional, although this is still under scrutiny. It is possible that HSPA7 could be a sometimes transcribed pseudogene. There is no conclusive evidence on whether its mRNA produces or not a protein with a defined function

Source Brocchieri L, Conway de Macario E, Macario AJL (2007) Chaperonomics, a new tool to study ageing and associated diseases. Mechan Ageing Develop. 128:125–136; and Brocchieri L, Conway de Macario E, Macario AJL (2008) *hsp-70* genes in the human genome: conservation and differentiation patterns predict a wide array of overlapping and specialized functions. BMC Evolutionary Biology 2008, 8:19; doi:10.1186/1471-2148-8-19. http://www.biomedcentral.com/1471-2148/8/19

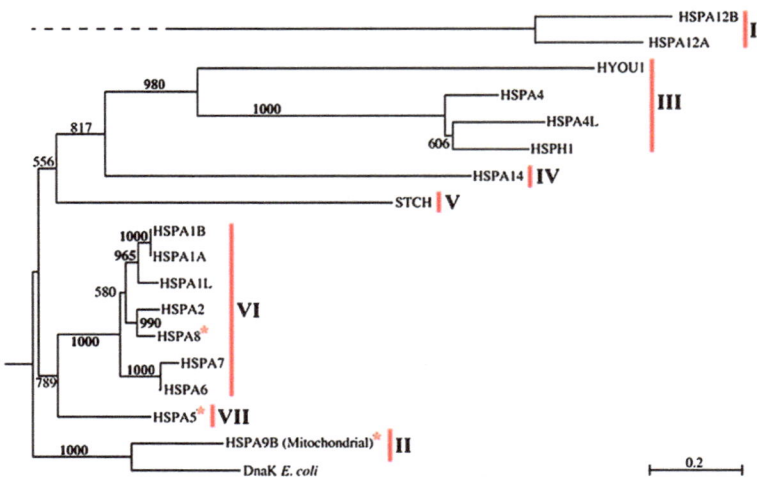

Fig. 2.4 A phylogenetic tree of the proteins encoded by the 17 *hsp70* genes distinguishing seven major evolutionarily related groups (shown by the *vertical red lines*), which were defined by bootstrap-support values over 85 %. The *asterisks* indicate the genes for which related pseudogenes were found. **Group I** was composed of the most diverged sequences, HSPA12A and HSPA12B, of uncertain relation with other eukaryotic and prokaryotic Hsp70s. Although conservation of a few sequence motifs clearly identifies these two genes as members of the extended *hsp70* gene family, their sequence conservation was insufficient to allow for a reliable, accurate determination of their evolutionary position compared to other human Hsp70s. **Group II** was composed of the mitochondrial protein HSPA9B, considered of alpha-proteobacterial origin, which accordingly, clustered with the DnaK sequence from *E. coli* (Gamma-proteobacteria) with very high bootstrap support. **Group III** encompassed the 105/110 kDa proteins HSPA4, HSPA4L, HSPH1 (clustered with 100 % bootstrap support) and the more distantly related sequence HYOU1, coding for the 170 kDa protein Grp170. HSPA14 was also related to proteins in Group III but with lower (81.7 %) bootstrap support than that (85 % or higher) adopted to identify closely related sequences as members of a distinct Group. Therefore, HSPA14 was classified separately in **Group IV**. Also joined to the sequences of Groups III and IV, was the sequence STCH, but with considerably lower bootstrap support (55.6 %) than that required to differentiate the Groups, and thus STCH was assigned to another group, i.e., **Group V**. **Group VI** was composed of sequences clustered with 100 % bootstrap support and, among them, three subgroups were distinguish, one including HSPA1A, HSPA1B and HSPA1L, a second including HSPA8 and HSPA2, and a third including HSPA6 and HSPA7. **Group VII** included sequence HSPA5, expressed in the endoplasmic-reticulum (*ER*), which was joined to Group VI with lower bootstrap support (78.9 %) than the minimum adopted for distinguishing the main groups. *Source* Brocchieri L, Conway de Macario E, Macario AJL (2008) *hsp-70* genes in the human genome: conservation and differentiation patterns predict a wide array of overlapping and specialized functions. BMC Evolutionary Biology 2008, 8:19; doi:10.1186/1471-2148-8-19 http://www.biomedcentral.com/1471-2148/8/19

Other genes related functionally with Hsp70. Information on another important group of Hsp-chaperones, Hsp40/DnaJ, can be found in Qiu XB, Shao YM, Miao S, Wang L (2006) The diversity of the DnaJ/Hsp40 family, the crucial partners for Hsp70 chaperones. Cell Mol Life Sci 63:2560–2570; and Kakkar V, Prins LC,

Table 2.6 *hsp90* genes

Group	Species studied	Total	HTPG (bacterial Hsp90)	TRAP (mito.) [HSPC5]a	HSP90C (chlor.)	HSP90B (ER) [HSPC4]	HSP90A (cytosolic) [HSPC1]	HSP90AA (cytosolic alpha) [HSPC2]	HSP90AB (cytosolic beta) [HSPC3]
Animalia	9	3–16	None (only in bacteria)	Yes (9 of 9 species)	None (only in plants)	Yes (all 9 species)	Yes (6 of 9 species)	Yes (3 of 9 species)	Yes (3 of 9 species)
Homo sapiens		5 (16)	None	1 (1)	None	1 (3)	None	2 (6)	1 (6)
All studied (6)	32								
Coding region exons (mRNA)		1–19	1 (1)–1 (1)	3 (3)–19–19	10 (?)–21 (21)	1 (?)–18 (18)	1 (1)–11 (12)	8 (?)–12 (12)	7 (7)–11 (12)
AA mature protein (precursor)		588–854 (588–854)	588 (588)–681 (681)	644 (688)–687 (719)	756 (810)–785 (811)	695 (711)–800 (823)	689 (689)–745 (745)	728 (728)–854 (854)	724 (724)–737 (737)
kDa mature protein (precursor)		66.7–98.1 (66.7–98.1)	66.7 (66.7)–78.0 (78.0)	74.8 (78.1)–77.9 (81.5)	84.2 (89.3)–89.0 (91.5)	79.5 (81.2)–91.5 (94.2)	78.3 (78.3)–86.2 (86.2)	84.1 (84.1)–98.1 (98.1)	83.3 (83.3)–84.8 (84.8)
Homo sapiens	Ref.		n.a	Ref.	n.a.	Ref.	n.a.	Ref.	Ref.

a *Abbreviations mito.*, mitochondria; *chlor.*, chloroplast; *ER*, endoplasmic reticulum; *Ref.*, see reference below; *n.a.*, not applicable. Within brackets, new nomenclature for the human molecules

Source Chen B, Piel WH, Gui L, Bruford E,Monteiro A (2005) The HSP90 family of genes in the human genome: Insights into their divergence and evolution. Genomics 86:627–637; Chen B, Zhong D, Monteiro A (2006) Comparative genomics and evolution of the HSP90 family of genes across all kingdoms of organisms. BMC Genomics 7:156 doi:10.1186/1471-2164-7-156; and Kampinga HH, Hageman J, Vos MJ, Kubota H, Tanguay RM, Bruford EA, Cheetham ME, Chen B, Hightower LE (2009) Guidelines for the nomenclature of the human heat shock proteins. Cell Stress Chaperones 14: 105–111; doi:10.1007/s12192-008-0068-7

Kampinga HH (2012) DNAJ proteins and protein aggregation diseases. Curr Top
Med Chem 12:2479–2490 (see also Sect. 4.3).

2.4.4 The Hsp90 Chaperone Genes and Proteins

The extended family of Hsp90 genes in the three life Domains comprises at least 17
genes, of which five are present in humans. These are: Hsp90AA1 (the cytosolic
Hsp89, or Hsp90, or HSPC1), Hsp90 alpha (or Hsp90AA2, or HSPC2), Hsp90 beta
(or HSP90B1, or HSPC3), TRAP (the mitochondrial Hsp90L, or Hsp75, or HSPC5),
and Hsp90B1 (the ER gp96, or grp94, or endoplasmin, or HSPC4). In addition, the
human genome harbors several related pseudogenes. The Hsp90 genes identified in
eukaryotes along with HTPG, which is present in bacteria, are listed in the Table 2.6.
These genes form four evolutionarily related groups (not shown).

2.5 The Chaperoning System

As a working hypothesis it may be assumed that all Hsp-chaperones in an
organism form a physiologic system, akin to the immune system. Thus, the
chaperoning system is composed of all Hsp-chaperones, co-chaperones and
chaperone co-factors of an organism. This is illustrated by the distribution and
migrations of certain chaperones, e.g., Hsp60, Fig. 2.5.

   The chaperone molecules can be classified into subpopulations according to their
origin with regard to the cell in which they function, precise location inside or
outside the cell, and to their mobility, and also considering whether or not they are
associated to other molecules or molecular assemblies. For example, according to its
origin a chaperone can be autochthonous if residing in the cell in which it originated
or imported if it comes from another cell; sessile if attached to a structure or mobile if
free-moving in the cytosol, extracellular environment or body fluids (blood, lymph,
cerebrospinal fluid), as illustrated in Fig. 2.5. In addition, chaperones can be
single or member of a chaperoning team with other chaperones, co-chaperones, and
co-factors, also named chaperoning machine (See Fig. 2.1). A chaperoning team or
machine can be a member of a chaperoning network, which is formed by various
chaperoning teams and, possibly, other molecules or molecular assemblies (See
Fig. 2.2). A chaperone can also form a complex with another molecule or structure
(e.g., tumor antigen, cell-surface receptor, cytosolic glucocorticoid-hormone
receptor, chemical compound, members of the caspase cascade), but in this case the
complex is not a chaperoning machine; it has other functions unrelated to its
canonical role in the maintenance of protein homeostasis. Examples of these non-
chaperoning complexes are: (1) Hsp70 forms complexes with tumor antigens
(peptides) and cell-surface receptors; it binds mRNAs containing AU-rich elements
and acts as posttranscriptional regulator of gene expression; HspA2 associates with a

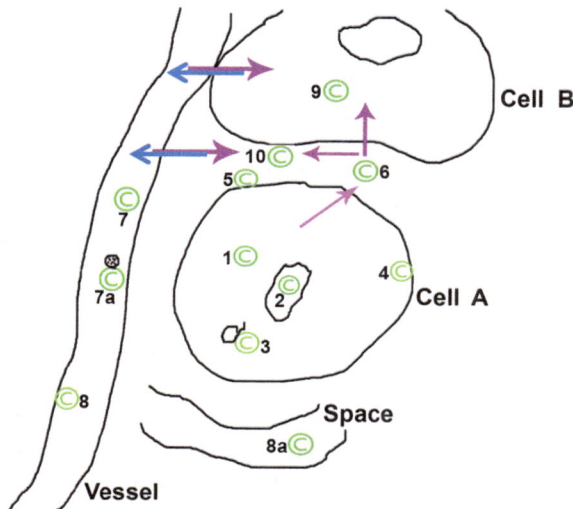

Fig. 2.5 The chaperoning system. *Circled C*, molecular chaperone; *1*, mobile chaperone in the cytosol; *2*, chaperone inside an organelle, such as the nucleus or a mitochondrion; *3*, sessile chaperone anchored to a particle (e.g., ribosome) in the cytosol; *4* and *5*, sessile chaperone anchored to the cell membrane on the cytosolic side (*4*) or on the outside in the extracellular space (*5*)–chaperones can also be located, at least transitorily, in the plasma membrane (i.e., transmembrane location); *6*, mobile chaperone in the intercellular space; *7*, mobile chaperone in circulation inside a vessel (blood or lymph) in suspension or, *7a*, on the surface of circulating erythrocytes, lymphocytes, granulocytes, or platelets; *8*, sessile chaperone anchored to the vessel wall on the inside; *8a*, chaperone inside a biological space, such as the intrasynovial space in the cavity of many joints, and the space between the pia and the arachnoid maters in the central nervous system (cerebrum ventricles, cisterns, and sulci, and spinal cord central canal); *9*, mobile chaperone in the cytosol like that shown in 1, but imported from another cell. Molecular chaperones can be found also in other locations such as cerebrospinal fluid (*8a*) and secretions (e.g., saliva and urine), the latter two not shown in this figure (see Table 2.2); *10*, mobile or sessile chaperone that originated in the blood or on a nearby cell (same as 6 if mobile) and is now in the intercellular space. *Arrows* indicate the various directions of movement of a chaperone molecule from inside a cell to the extracellular space or vessel lumen and vice versa. A chaperone can gain the extracellular space from inside a cell or from inside a vessel and it can go into a vessel directly from a cell or from the extracellular space. *Source* Macario AJL, Conway de Macario E (2009) The chaperoning system: Physiology and pathology. Exp Med Rev Vol. 2–3: Years 2008/09, pp. 9–21 (http://www.unipa.it/giovanni.zummo); and Macario AJL, Cappello F, Zummo G, Conway de Macario E (2010) Chaperonopathies of senescence and the scrambling of the interactions between the chaperoning and the immune systems. Ann New York Acad Sci 1197: 85–93

complex including arylsulfatase A (ARSA) and sperm adhesion molecule 1 (SPAM1) in the process of spermatozoa capacitation; (2) Hsp90 binds glucocorti-coid-hormone receptor (a protein that is a transcription factor); (3) Hsp90 binds some anti-tumor compounds like the antibiotic geldanamycin; and (4) Hsp60 forms a complex with procaspase-3.

The eukaryotic cell, for example a cell in the human body, has organelles, such as mitochondria and the endoplasmic reticulum (ER). These organelles have their own subsets of chaperones, i.e., chaperoning subsystems, with defined functions in health and roles in pathology. For instance, the mitochondrial chaperoning subsystem is implicated in the pathogenesis of various conditions such as ageing, and some types of cancer and neurological disorders (e.g., mitochondrial encephalopathy). In these conditions, mitochondrial functions are abnormal. Likewise, the ER chaperoning subsystem plays a pathogenic role in some malignant tumors, and genetic (i.e., the Marinesco-Sjogren syndrome) and inflammatory disorders (e.g., synovial inflammation in some forms of arthritis), just to mention a few examples. See Table 4.9, part 1; Table 7.2; and Table 7.4.

The concept of chaperoning system offers a unified view of a set of molecules that interact directly or indirectly with one another to achieve a particular objective, i.e., to carry out a physiological process usually essential for life and in the response to stress. It may be assumed that the chaperoning system, or very primitive forms of it, appeared very early in evolution and played a defensive role against stressors. With time, the system became multifacetic and it assumed also a regulatory role as the living forms became more complex, multicellular and multiorgan, therefore needing internal coordination of parts. This role could naturally be played by a population of circulating molecules such as the Hsp-chaperones able to interact widely, including with the immune system. The latter probably appeared later in evolution to assume also defensive roles against a widening range of stressors, including microbes and eukaryotic parasites. The result is that today, the chaperoning and the immune systems interact widely.

From the practical standpoint the concept of chaperoning system is a working hypothesis that provides a scaffolding to add pertinent data as they become available and a blueprint for future research to obtain new data. It also serves as a guide for diagnosis and differential diagnosis, and for assessing the impact throughout the body that any given chaperonopathy might have. All of these are essential requisites for adequate treatment and prevention.

Further Reading

See also Chap. 1 for Further Reading

Sections 2.1–2.4

Clusterin

Carver JA, Rekas A, Thorn DC, Wilson MR (2003) Small heat-shock proteins and clusterings: intra- and extracellular molecular chaperones with a common mechanism of action and function? IUBMB Life 55:661–668

Wyatt AR, Yerbury JJ, Berghofer P, Greguric I, Katsifis A, Dobson CM, Wilson MR (2011) Clusterin facilitates in vivo clearance of extracellular misfolded proteins. Cell Mol Life Sci 68:3919–3931. doi:10.1007/s00018-011-0684-8

Alpha Synuclein and Myocilin

Anderssohn AM, Cox K, O'Malley K, Dees S, Hosseini M, Boren L, Wagner A, Bradley JM, Kelley MJ, Acott TS (2011) Molecular chaperone function for myocilin. Invest Ophthalmol Vis Sci 52:7548–7555

Rekas A, Ahn KJ, Kim J, Carver JA (2012) The chaperone activity of α-synuclein: Utilizing deletion mutants to map its interaction with target proteins. Proteins 80:1316–1325

AIPL1 Part of Chaperone Complex

AIPL1 (aryl hydrocarbon-interacting receptor protein-like 1)

Hidalgo-de-Quintana J, Evans RJ, Cheetham ME, van der Spuy J (2008) The Leber congenital amaurosis protein AIPL1 functions as part of a chaperone heterocomplex. Invest Ophthalmol Vis Sci 49:2878–2887

AIP

AIP (aryl-hydrocarbon receptor-interacting protein; or aryl hydrocarbon receptor-associated protein 9, ARA9)

Cain JW, Miljic D, Popovic V, Korbonits M (2010) Role of the aryl hydrocarbon receptor-interacting protein in familial isolated pituitary adenoma. Experts Rev Endocrinol Metabolism 5:681–695

Torsin

Burdette AJ, Churchill PF, Caldwell GA, Caldwell KA (2010) The early-onset torsion dystonia-associated protein, torsinA, displays molecular chaperone activity in vitro. Cell Stress Chaperones 15:605–617

Chen P, Burdette AJ, Porter JC, Ricketts JC, Fox SA, Nery FC, Hewett JW, Berkowitz LA, Breakefield XO, Caldwell KA, Caldwell GA (2010) The early-onset torsion dystonia-associated protein, torsinA, is a homeostatic regulator of endoplasmic reticulum stress response. Hum Mol Genet 19:3502–3515

Ozelius LJ, Page CE, Klein C, Hewett JW, Mineta M, Leung J, Shalish C, Bressman SB, de Leon D, Brin MF, Fahn S, Corey DP, Breakefield XO (1999) The TOR1A (DYT1) gene family and its role in early onset torsion dystonia. Genomics 62:377–384

AHSP (Alpha-Hemoglobin-Stabilizing Protein)

Bank A (2007) AHSP: a novel hemoglobin helper. J Clin Invest 117:1746–1749

Favero ME (2011) Costa FF (2011) alpha-hemoglobin-stabilizing protein: an erythroid molecular chaperone. Biochem Res Int 2011:373859

Hsp10

Corrao S, Campanella C, Anzalone R, Farina F, Zummo G, Conway de Macario E, Macario AJL, Cappello F, La Rocca G (2010) Human Hsp10 and early pregnancy factor (EPF) and their relationship and involvement in cancer and immunity: current knowledge and perspectives. Life Sci 86:145–152

Czarnecka AM, Campanella C, Zummo G, Cappello F (2006) Heat shock protein 10 and signal transduction: a "capsula eburnea" of carcinogenesis? Cell Stress Chaperones 11:287–294

Chaperoning Networks: Balance Between Protein Folding and Degradation

Muller P, Ruckova E, Halada P, Coates PJ, Hrstka R, Lane DP, Vojtesek B (2012) C-terminal
    phosphorylation of Hsp70 and Hsp90 regulates alternate binding to co-chaperones CHIP and
    HOP to determine cellular protein folding/degradation balances. Oncogene. doi:10.1038/
    onc.2012.314

CHAPERONES: REALM
CCT, The Cytosolic Chaperonin is also Present in the Nucleus and Perform
Functions Other than the Canonical Protein Folding

Pejanovic N, Hochrainer K, Liu T, Aerne BL, Soares MP, Anrather J (2012) Regulation of
    nuclear factor κB (NF–κB) transcriptional activity via p65 acetylation by the chaperonin
    containing TCP1 (CCT). PLoS ONE 7(7):e42020
Huang R, Yu M, Li CY, Zhan YQ, Xu WX, Xu F, Ge CH, Li W, Yang XM (2012) New insights
    into the functions and localization of nuclear CCT protein complex in K562 leukemia cells.
    Proteomics Clin Appl 6:467–475

Chaperones on the Surface of Tumor Cells

Stangl S, Gehrmann M, Riegger J, Kuhs K, Riederer I, Sievert W, Hube K, Mocikat R, Dressel R,
    Kremmer E, Pockley AG, Friedrich L, Vigh L, Skerra A, Multhoff G (2011) Targeting
    membrane heat-shock protein 70 (Hsp70) on tumors by cmHsp70.1 antibody. Proc Natl Acad
    Sci USA 108:733–738
Shipp C, Derhovanessian E, Pawelec G (2012) Effect of culture at low oxygen tension on the
    expression of heat shock proteins in a panel of melanoma cell lines. PLoS ONE 7(6):e37475

Saliva and Serum Hsp60

Yuan J, Dunn P, Martinus RD (2011) Detection of Hsp60 in saliva and serum from type 2
    diabetic and non-diabetic control subjects. Cell Stress Chaperones 16:689–693

ER chaperones are Also Outside the Organelle

Gold LI, Eggleton P, Sweetwyne MT, Van Duyn LB, Greives MR, Naylor SM, Michalak M,
    Murphy-Ullrich JE (2010) Calreticulin: non-endoplasmic reticulum functions in physiology
    and disease. FASEB J 24:665–683

How Many Chaperones in the Human Species?

Brocchieri L, Conway de Macario E, Macario AJL (2007) Chaperonomics, a new tool to study
    ageing and associated diseases. Mechan Ageing Develop 128:125–136

Section 2.5

The Chaperonin System and Non-chaperoning Complexes Formed by
Chaperones (see also Further Reading in Sects. 7.3–7.4)

Csermely P, Korcsmáros T, Kovács IA, Szalay MS, Soti C (2008) Systems biology of molecular
    chaperone networks. Novartis Found Symp 291:45–54; discussion 54–8, 137–40

Echeverría PC, Bernthaler A, Dupuis P, Mayer B, Picard D (2011) An interaction network predicted from public data as a discovery tool: application to the Hsp90 molecular chaperone machine. PLoS ONE. 2011; 6(10): e2604410.1371/journal.pone.0026044

Kishor A, Tandukar B, Ly YV, Toth EA, Suarez Y, Brewer G, Wilson GM (2013) Hsp70 is a novel posttranscriptional regulator of gene expression that binds and stabilizes selected mRNAs containing AU-rich elements. Mol Cell Biol 33:71–84. doi:10.1128/MCB.01275-12

Redgrove KA, Anderson AL, McLaughlin EA, O'Bryan MK, Aitken RJ, Nixon B (2013) Investigation of the mechanisms by which the molecular chaperone HSPA2 regulates the expression of sperm surface receptors involved in human sperm-oocyte recognition. Mol Hum Reprod 19:120–135. doi:10.1093/molehr/gas064

Chapter 3
The Chaperonopathies: Classification, Mechanisms, Structural Features

Abstract The classification of chaperonopathies is presented in this chapter. Like many other diseases, chaperonopathies can be genetic or acquired, primary or secondary, structural and/or functional, and qualitative and/or quantitative. In addition, considering pathogenic mechanism, chaperonopathies can be by defect, excess, or mistake. In the latter, a chaperone is normal but favors disease, a situation that occurs, for instance, in various types of cancers. Structural chaperonopathies are characterized by a change in the molecule of a chaperone due to mutation (genetic chaperonopathy) or due to aberrant post-translational modification (acquired chaperonopathy). In both cases, the impact of the structural change depends on which functional domain within the chaperone molecule is modified.

Keywords Chaperonopathies, classification · Chaperonopathies, by defect · Chaperonopathies, by excess · Chaperonopathies, by mistake · Chaperone, structural-functional domain · Primary chaperonopathies · Proteinopathies · Proteotoxicity · Secondary chaperonopathies · Chaperonin disease · Chaperonin pathology · Sick chaperones

3.1 Classification of Chaperonopathies

Chaperonopathies can be classified in various ways as many other groups of diseases: for example, considering the mechanism responsible for the chaperone abnormality, chaperonopathies can be distinguished into genetic, epigenetic, and acquired. Likewise, considering whether the chaperone molecule has a structural change in itself that affects its function, or its function is affected by another molecule, chaperonopathies can be sorted out into structural and functional. Taking into consideration their role in disease, chaperonopathies can be primary (they are *the, or one*, etiological-pathogenic factor) and secondary (they are the manifestation of another disease, which is not a chaperonopathy in itself). In addition, bearing in mind the way they manifest themselves and cause pathology, chaperonopathies can be classified into by excess, defect, or mistake, considering

Table 3.1 Classification of chaperonopathies according to pathogenic mechanism

Chaperonopathies by:	Mechanism, features
Excess	Quantitative, e.g., due to gene dysregulation and overexpression
	Qualitative, e.g., gain of function
Defect	Quantitative, e.g., gene downregulation
	Qualitative, e.g., due to structural defect genetic or acquired
Mistake	Normal chaperones can contribute to disease, e.g., some tumors that need chaperones to grow; autoimmune conditions in which a chaperone is the autoantigen; and possibly also prion diseases in which chaperones may be required for propagation

Source Macario AJL, Conway de Macario E (2007) Chaperonopathies by defect, excess, or mistake. Ann N Y Acad Sci 1113:178–191; and Cappello F, Macario AJL (2012) "Mitochondrial chaperonin Hsp60: locations, functions and pathology," in Houry, W.A. (ed.), *Protein Homeostasis:* The Biomedical & Life Sciences Collection, Henry Stewart Talks Ltd, London (online at http://hstalks.com/bio). **Direct talk access links:** http://hstalks.com/lib.php?t=HST142.3142&c=252

the quantity (e.g., concentration in biological fluids, and density in cells and tissues) of the affected chaperone and/or its level and type of function (e.g., hypofunctional, dysfunctional, and gain of function) (Table 3.1).

Quantitative variations in chaperone levels may be due to various mechanisms. For example, an increase of a chaperone can be due to gene overexpression, diminished degradation of its mRNA or of the protein itself, and other factors such as increased efficiency of translation, and slower trafficking than normal through any given cell compartment. It is likely that a combination of mechanisms contribute to augment the levels of one or more chaperones in a cell. This situation may happen if there is increased demand for chaperones, in the cytosol for instance, by accumulating abnormal polypeptides in disorders such as the proteins filamin in filaminopathies, tau in tauopathies (for instance Alzheimer's disease, progressive supranuclear palsy, frontotemporal dementia with parkinsonism linked to chromosome 17, Lytico-Bodig disease, Pick's disease, corticobasal degeneration, Hallevorden-Spatz disease, and others—there are at least 15 of these pathological conditions), alpha-synuclein in Parkinson's disease, or proteins with a polyglutamine (PolyQ) expansion (e.g., huntingtin in Huntington's disease, androgen receptor in spinal and bulbar muscular atrophy, and others such as ataxin in some spinocerebellar ataxias). But in these cases, a decrease in the levels of one or more chaperones might be observed at one time or another instead of an increase. It may be assumed that the observed decrease of chaperones in the cytosol is the result of their becoming trapped in the fibrillary tangles, Lewy bodies, or other kinds of protein aggregates and, thus, are no longer present with their normal distribution.

We will not discuss in this book the so called "amateur" pathological chaperones, e.g., apolipoproteins and heparan sulfate proteoglycans. Apolipoprotein E (ApoE) and alpha-1-anti-chymiotrypsin (ACT) have been found augmented in some instances of Alzheimer's disease, and implicated in the formation of fibrillary precipitates of Beta-amyloid in senile plaques.

Likewise, we will not review the experimental work on the role of chaperones in prion propagation and biology. This issue is still under investigation in human diseases using human materials. This group of pathological conditions sometimes designated transmissible spongiform diseases or transmissible spongiform encephalopathies, encompass Creutzfeldt-Jakob disease (CJD) with its various forms, iatrogenic, variant, familial, and sporadic; Gerstmann-Sträussler-Scheinker syndrome (GSS); fatal familial insomnia (FFI); and kuru. The role of molecular chaperones in the pathogenesis of these diseases is suspected to be significant from the information obtained with experimental systems, such as baker's yeast (*Saccharomyces cerevisiae*). This is a field that deserves investigation since elucidation of the role of chaperones in the mechanism of disease development and transmission could lead to advances in prevention, diagnosis, and treatment.

It is pertinent to ask the question whether in pathologies showing quantitative variations of chaperones the alterations observed are cause or consequence of disease, namely if they are primary or secondary chaperonopathies. In any case, they must be scrutinized because if the quantitative changes entail a participation in pathogenesis by excess or defect, the diseases would be genuine chaperonopathies. If the quantitative changes simply are the consequences of another disorder, still they can be considered markers of disease (secondary chaperonopathies), with potential in diagnosis, disease monitoring, and assessment of response to treatment. Cases in point are certain types of cancer that show increased levels of chaperones (see Table 7.2). One may envision various mechanisms underlying the increase of chaperones in cancer. For example, cancer cells usually have a very high rate of protein synthesis with production of large amounts of polypeptides in need of assistance from chaperones for folding, which in turn would result in an increment of the chaperones inside the cell. In addition, it has been observed in cancer cells that not only the quantity of chaperones is elevated but also their distribution is altered with regard to normal cells. For instance, in colon cancer cells Hsp60 is augmented and appears not only inside the mitochondria, its typical residence, but also in the cytosol and in the plasma-cell membrane (similar patterns have been observed in other types of cancer and for other chaperones). It is tempting to speculate that this abundance and re-distribution of Hsp60 (or any other chaperone) is not simply overflow from intense production but it may represent either a defensive mechanism implemented by the normal cells undergoing malignant transformation, or a way that the tumor engineers to favor its own cause by, for instance, sending emissaries (the increased chaperones that actively secreted by the cancer cells) to other cells in the vicinity, or far away via circulation. These and other alternatives, such as the presence and cause of post-translational modifications of chaperones in cancer and other diseases that drive the chaperones to perform functions that they do not normally do (e.g., promote tumor-cell growth, survival and migration), are the subject of current research that should provide important information on chaperones, chaperonopathies, and cancer (see Chap. 9).

Chaperonopathies can be caused by one or more of various factors, e.g., mutation of a chaperone gene or post-translational modification of the chaperone molecule. In addition, it is possible that epigenetic mechanisms might be involved in the causation and perpetuation of some chaperonopathies. It is well established that certain phenotypes are due to changes in gene expression without changes in the nucleotide sequence, e.g., mutation, but associated with other types of DNA modifications, e.g., methylation. Likewise, histone modifications (e.g., acetylation) can alter gene expression without altering the nucleotide sequence. A modulation of a chaperone gene by modified histones could very well result in certain plasticity in gene expression. If we bear this idea in mind, our understanding of chaperonopathies could probably be advanced considerably in previously unsuspected directions. We know that many of these processes due to modified DNA (without changes in the nucleotide sequence), or modified histones, can be passed to descendants, are reversible, and can occur at any age. This mechanism could be implicated in certain seemingly genetic conditions whose familial incidence cannot be explained by the canonical Mendelian principles.

The probability that epigenetic mechanisms play a key role in the biology and pathology of at least some chaperones is a priori high, considering that chaperones are essential players in the response of the cell and the organism to environmental changes. Thus, epigenetic mechanisms should be keenly investigated in those cases of chaperonopathies that seem inheritable but that do not satisfy classical genetic rules. There is already some information in the literature indicating that epigenetic mechanisms might be implicated in chaperonopathies. For instance, the clusterin gene was found methylated at the promoter and downregulated in normal breast tissue in which clusterin levels were low. In contrast, in some cases of highly malignant breast cancer, the clusterin-gene promoter was unmethylated and overexpressed. This would be an example of chaperonopathy by mistake or collaborationism with the culprit being the chaperone clusterin (see Sect. 7.2). Likewise, hypermethylation of the promoter of the alpha-Crystallin gene CRYAA (see Table 2.3) was found to decrease the expression of this gene and it was observed that this epigenetic repression of the CRYAA gene in the lens epithelium was implicated in the pathogenesis of age-related nuclear cataract. More recently the Cosmc gene promoter was found methylated and, thereby, silenced in a B-cell line derived from a patient with the Tn-syndrome (see Table 4.9, part 2), resulting in abnormal expression of the Tn antigen, which is thought to be pathogenic.

3.2 Primary Chaperonopathies

Primary chaperonopathies are those caused by mutation of the affected chaperone gene and can, in principle, be corrected by chaperonotherapy (Fig. 3.1).

```
┌──────────────────────────────────────────────────┐
│  PRIMARY CHAPERONOPATHIES                          │
├──────────────────────────────────────────────────┤
│  Gene mutation                                     │
│                                                    │
│  Hereditary                                        │
│                                                    │
│  Congenital                                        │
│                                                    │
│  Complex phenotype. Ageing fast                    │
│                                                    │
│  Normal chaperones to the rescue                   │
│                                                    │
│  Chaperonotherapy: gene and/or protein             │
└──────────────────────────────────────────────────┘
```

Fig. 3.1 Primary chaperonopathies are caused, for example, by a mutation in a chaperone gene. They are hereditary, and usually of early onset, with a complex phenotype, most likely reflecting the wide distribution of the affected chaperone in tissues and cells, in which the occurrence of the defective chaperone has a negative impact (see Sect. 2.3). *Source* Macario AJL, Conway de Macario E (2007) Chaperonopathies and chaperonotherapy. FEBS Lett 581:3681–3688

3.3 Structural Features of Genetic Chaperonopathies

Typically, chaperone molecules are made of various distinct domains with specific functions. A mutation (or other structural alteration in a domain) may cause a chaperonopathy, whose manifestations will depend on the functions of the domain affected (Fig. 3.2). See also Sect. 2.2.

Not all chaperone molecules have all the domains represented in the figure and the domains are not necessarily organized in the sequence shown. For instance, some chaperones do not have an ATP-binding site or ATPase ability, in which case the energy is provided by a co-chaperone (i.e., another member of the chaperoning team that have both).

Examples of structural hereditary chaperonopathies affecting the various families-types of Hsp-chaperones are given in the following pages. It has to be borne in mind that in several genetic chaperonopathies more than one gene, not just a chaperone gene, have been found mutated, indicating that they are polygenic, heterogenous conditions.

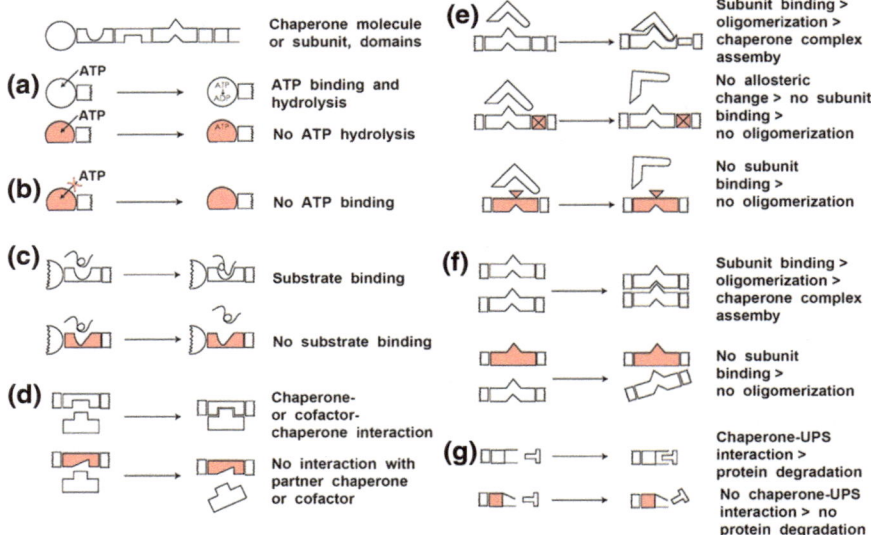

Fig. 3.2 Key for the structural domains, from left to right in top left scheme: ATP-binding-ATPase, substrate binding, chaperone or cofactor-chaperone interaction (needed for the assembly of the chaperoning networks, i.e., interaction with other chaperones and chaperoning teams), oligomerization (needed for the formation of oligomers, i.e., the chaperoning complex or team such as the homo-heptamer formed by Hsp60 or the hetero-octamer characteristic of CCT); hinge (needed for allowing the allosteric changes accompanying all functions of the chaperone molecule—several of these domains are usually present); ubiquitin-proteasome interaction (needed for interaction with the ubiquitin-proteasome system for protein degradation). Filled forms represent structurally altered domains due to mutation or post-translational modification. UPS, ubiquitin-proteasome system. *Source* Macario AJL, Conway de Macario E (2007) Molecular chaperones: Multiple functions, pathologies, and potential applications. Front Biosci 12: 2588–2600. To see Table of Contents, Abstract, Figures, and Tables go to http://www.bioscience. org/2007/v12/af/2257/fulltext.htm

Further Reading

See also Chap. 2 for Further Reading

Sections 3.1–3.3

Chaperonopathies vs. Proteinopathies

Macario AJL, Conway de Macario E (2000) Stress and molecular chaperones in disease. Intl J Clin Lab Res (Currently J Clin Exp Med; Springer) 30:49–66

PROTEINOPATHIES
Quantitative Changes in Chaperones and Chaperones in Pathological Protein Precipitates
Macario AJL, Conway de Macario E (2001) Molecular chaperones and age-related degenerative disorders. Adv Cell Aging Gerontol 7:131–162

Proteinopathies and Proteotoxicity
Willis MS, Patterson C (2013) Proteotoxicity and cardiac dysfunction–Alzheimer's disease of the heart? N Engl J Med 368:455–464. doi:10.1056/NEJMra1106180

Filaminopathies
Kley RA, van der Ven PF, Olivé M, Höhfeld J, Goldfarb LG, Fürst DO, Vorgerd M (2013) Impairment of protein degradation in myofibrillar myopathy caused by FLNC/filamin C mutations. Autophagy 9:422–423. doi:10.4161/auto.22921

Tauopathies, for instance Alzheimer's disease
Thompson AD, Scaglione KM, Prensner J, Gillies AT, Chinnaiyan A, Paulson HL, Jinwal U, Dickey CA, Gestwicki JE (2012) Analysis of the tau-associated proteome reveals that exchange of Hsp70 for Hsp90 is involved in tau degradation. ACS Chem Biol 7:1677–1686
Miyata Y, Koren J, Kiray J, Dickey CA, Gestwicki JE (2011) Molecular chaperones and regulation of tau quality control: strategies for drug discovery in tauopathies. Future Med Chem 3:1523–1537

Alpha-Synucleinopathies: For Instance Parkinson's Disease
Dimant H, Ebrahimi-Fakhari D, McLean PJ (2012) Molecular chaperones and co-chaperones in Parkinson disease. Neuroscientist 18:589–601
Redeker V, Pemberton S, Bienvenut W, Bousset L, Melki R (2012) Identification of protein interfaces between alpha-synuclein, the principal component of Lewy bodies in Parkinson's disease, and the molecular chaperones human Hsc70 and the yeast Ssa1p. J Biol Chem 287:32630–32639

Polyglutamine Expansion Diseases: For Instance Huntington's Disease and Some Ataxias
Kubota H, Kitamura A, Nagata K (2011) Analyzing the aggregation of polyglutamine-expansion proteins and its modulation by molecular chaperones. Methods 53:267–274
Trott A, Houenou LJ (2012) Mini-review: spinocerebellar ataxias: an update of SCA genes. Recent Pat DNA Gene Seq 6:115–121
Waelter S, Boeddrich A, Lurz R, Scherzinger E, Lueder G, Lehrach H, Wanker EE (2012) Accumulation of mutant huntingtin fragments in aggresome-like inclusion bodies as a result of insufficient protein degradation. Mol Biol Cell 12:1393–1407

Prion Diseases
Collins S, McLean CA, Masters CL (2001) Gerstmann-Sträussler-Scheinker syndrome, fatal familial insomnia, and kuru: a review of these less common human transmissible spongiform encephalopathies. J Clin Neurosci 8:387–397

Jones GW, Tuite MF (2005) Chaperoning prions: the cellular machinery for propagating an infectious protein? Bio Essays 27:823–832
Kiktev DA, Patterson JC, Müller S, Bariar B, Pan T, Chernoff YO (2012) Regulation of chaperone effects on a yeast prion by cochaperone sgt2. Mol Cell Biol 32:4960–4970
Masison DC, Kirkland PA, Sharma D (2009) Influence of Hsp70s and their regulators on yeast prion propagation. Prion 3:65–73
True HL (2006) The battle of the fold: chaperones take on prions. Trends Genet 22:110–117

PATHOLOGICAL "AMATEUR" CHAPERONES

Potter H, Wefes IM, Nilsson LNG (2001) The inflammation-induced pathological chaperones ACT and apo-E are necessary catalysts of Alzheimer amyloid formation. Neurobiol Aging 22:923–930
Wilhelmus MM, de Waal RM, Verbeek MM (2007) Heat shock proteins and amateur chaperones in amyloid-Beta accumulation and clearance in Alzheimer's disease. Mol Neurobiol 35:203–216

POST-TRANSLATIONAL MODIFICATIONS OF CHAPERONES AND IMPACT ON FUNCTION

Rao R, Fiskus W, Ganguly S, Kambhampati S, Bhalla KN (2012) HDAC inhibitors and chaperone function. Adv Cancer Res 116:239–262

Chapter 4
Structural and Hereditary Chaperonopathies: Mutation

Abstract This chapter deals with structural and hereditary chaperonopathies. The chaperonopathies caused by mutations in: sHsp, chaperonin genes (Hsp60 or Cpn60, and CCT subunits), Hsp40/DnaJ, Hsp70, sacsin, and dedicated chaperones (e.g., those involved in microtubule biogenesis, in maintenance of the respiratory chain inside the mitochondria, and others in various cell compartments and tissues), are described and discussed.

Keywords α-crystallin · Cataracts · Myopathies · Distal neuropathies · Cardiopathies · Spastic paraplegias · Bardet-Biedl syndrome · Muscular dystrophies · Kufs disease · Parkinson · Sacsinopathies · Dystonias · Anemia · Chaperonin pathology · Abnormal chaperonins

Mutations in chaperone genes have been found associated with diseases and syndromes, examples of which are given in the following tables. It has to be borne in mind that in many pathological conditions more than one gene may be mutated, thus, the conditions are polygenic. This is the case for some of the diseases and syndromes listed in this chapter's tables. In chaperonopathies, the only gene mutated, or at least one of the genes mutated, encodes a protein known to be a chaperone or that it is similar and evolutionarily related to a well characterized chaperone. A case in point is Bardet-Biedl Syndrome (BBS). This is a typical genetically heterogeneous and pleiotropic disorder in which at least 14 genes (BBS genes) have been implicated. One group of these genes, BBS6 or McKusick-Kaufman Syndrome (MKKS), BBS10, and BBS12 encode proteins evolutionarily related to a well characterized chaperone, CCT8 (see Fig. 2.3). Other BBS genes encode proteins that are not chaperones or related to chaperones and form a complex, the BBSome, involved in ciliogenesis. Interestingly, BBS6, BBS10, and BBS12 may assist in a chaperoning-like fashion the assemblage of the BBSome. Patients with BBS may have one or more BBS genes mutated in various combinations; in some patients none of the BBS chaperones genes is mutated and, consequently, these BBS cases are not chaperonopathies.

Therefore, any given syndrome discussed in this book as genetic chaperonopathy would be a chaperonopathy only when a chaperone gene is found mutated and associated with pathology. In other words, the clinical-pathological diagnosis of BBS, for example, does not mean it is a chaperonopathy until a genetic analysis has proven that one or more of the three BBS chaperone genes is abnormal and associated with the condition. Likewise, osteogenesis imperfecta (brittle bone disease) is a term that encompasses hereditary connective tissue disorders characterized by defects in Type I collagen and alterations of its biosynthesis causing bone brittleness. Most of these disorders are associated with mutations in collagen genes but a minority of cases show mutation in chaperones assigned to collagen assembly. Only the latter cases of osteogenesis imperfecta can be considered candidates to be chaperonopathies. These considerations apply to many of the other diseases listed in the following tables, such as myopathies, distal neuropathies, cardiomyopathies, and so on. When any of these diagnoses has been concluded, the next step should be to elucidate the etiologic-pathogenic factor(s) causing the condition and determine if it is associated with a chaperonopathy. We would like to repeat here what we said earlier (Sect. 1.4): 'We consider that a diagnostic exercise does not conclude until the etiology of the syndrome under examination has been elucidated, and the pathogenesis, i.e., the mechanism by which the causal factor (the etiology) causes lesions and disease, has been understood. In our view, diagnosis is not just the labeling of a patient with the name of a known syndrome but it includes also the elucidation of etiology and pathogenesis. *Here, lies the importance of learning about chaperonopathies.* They are the etiologic-pathogenic factors in a variety of diseases as already established, and seem to be also implicated in many others. The latter require the special attention of, and diagnostic efforts by, clinicians and pathologists so the role of chaperonopathies in their pathogenesis can be confirmed or ruled out. One of the objectives of this book is to provide information to clinicians and pathologists so they can identify probable chaperonopathies and proceed to study them to confirm or rule out their initial suspicion'.

4.1 Chaperonopathies Due to Mutations in Small-Size Chaperones

Examples of chaperonopathies due to mutations in the genes of chaperones with a MW 34 kDa or less (see Table 2.1) are listed in Table 4.1.

Chaperonopathies associated with defects in members of the sHsp, crystallin subgroup (see Tables 2.1 and 2.3), tend to affect the eye lens, voluntary muscles, and peripheral nerves. Other chaperones exist with MW within the range of those

Table 4.1 Examples of structural-hereditary chaperonopathies affecting chaperones of MW of 34 kDa or less

Gene/protein affected	Disease/syndrome
Hsp27 (HspB1)	Williams
	Charcot-Marie-Tooth (CTM)
	Distal hereditary motor neuropathy (DHMN)
AlphaA and gammaC crystallins (HspB4; CRYAA; and CRYGC)	Childhood cataracts
AlphaB-crystallin (HspB5; CRYAB)	Desmin-related myopathy; other
Hsp22 (HspB8; CRYAC)	CMT 2L; DHMN IIA
PPI (peptidyl-prolil *cis-trans* isomerase); FKBP10; FKBP14	Osteogenesis imperfecta (recessive Bruck); Ehlers-Danlos (recessive EDS)
AIPL1 (aryl hydrocarbon-interacting receptor protein-like 1)	Leber congenital amaurosis (LCA)
AIP (aryl-hydrocarbon receptor-interacting protein)	Pituitary adenoma predisposition (PAP)

Source Macario AJL, Grippo TM, Conway de Macario E (2005) Genetic disorders involving molecular-chaperone genes: A perspective. Genet Med 7:3–12; and Macario AJL, Conway de Macario E (2005) Sick chaperones, cellular stress and disease. New Eng J Med 353:1489–1501; Macario AJL, Conway de Macario E (2007) Chaperonopathies by defect, excess, or mistake. Ann NY Acad Sci 1113:178–191; and the following Databases: Hugo Gene Nomenclature Committee (HGNC; http://www.genenames.org/); Online Mendelian Inheritance in Man (OMIN; http://omim.org/); UniProt (http://www.uniprot.org/); National Center for Biotechnology Information (NCBI; http://www.ncbi.nlm.nih.gov/guide/); and PubMed-NCBI (http://www.ncbi.nlm.nih.gov/pubmed)

of the sHsp crystallin family but do not belong to this family because they do not have the crystallin domain (see Table 2.1); chaperonopathies pertaining to these non-crystalin sHsp can affect various tissues and organs. For example, PPI chaperonopathies have the most impact on collagen-rich tissues. AIPL1 is associated with blindness and AIP with a predisposition to develop pituitary adenomas and gigantism. Examples of mutations and associated syndromes are listed in Table 4.2.

4.2   Chaperonopathies Due to Mutations in the Chaperonin Genes

Mutations in the genes encoding the chaperonins of Group I, i.e., Hsp60 or Cpn60, and the chaperonins of Group II, i.e., subunits of CCT (see Table 2.4), can cause disease, Tables 4.3, 4.4, 4.5 and 4.6.

Table 4.2 Human sHsp-associated chaperonopathies: examples of mutation

Name/HGNC/Gene ID/UniProtKB/ Swiss-Prot	Synonyms	Total aa	Chaperonopathies: Mutations
CRYAA; crystallin, alpha A/2388/1409/P02489	CRYA1; HSPB4; HspB4	173	Cataracts: arg116-to-cys (**R116C**); arg116-to-his (**R116H**); trp9-to-ter (**W9X**); arg49-to-cys (**R49C**)
CRYAB; crystallin alpha B/2389/1410/P02511	CRYA2; HSPB5; HspB5	175	Myopathies; Desmin-related myopathy with lens opacity: gln151-to-ter (**Q151X**) (truncated protein of 150 aa); arg157-to-his (**R157H**), Cataracts: arg120-to-gly (**R120G**). Multisystemic: asp109-to-his (**D109H**)
CRYAC; heat shock 22 kDa protein 8/30171/26353/ Q9UJY1	HSPB8; HSP22; HspB8	196	Charcot-Marie-Tooth (CMT) Type 2L; Distal Hereditary Motor Neuropathies (DHMN) Type IIA: lys141-to-asn (**K141N**); lys141-to-glu (**K141E**)
HSPB1; heat shock 27 kDa protein 1/5246/3315/P04792	HSP27; HSP28; HspB1	205	CMT, DHMN: arg127-to-trp (**R127W**); ser135-to-phe (**S135F**); arg136-to-trp (**R136W**); lys141 to gln (**K141Q**); thr151-to-ile (**T151I**); pro182-to-leu (**P182L**); pro182-to-ser (**P182S**)
HSPB6; heat shock protein alpha-crystallin-related B6/26511/ 126393/O14558	FLJ32389; Hsp20; HspB6	160	Cardiopathy; pro20-to-leu (**P20L**): decreased phosphorylation at Ser 16 and loss of anti-apoptotic action of mutant Hsp20
ODF1; outer dense fiber of sperm tails 1// 8113/4956/Q14990	HSPB10; HspB10	250	No known pathology (infertility ?)

Source The Databases: Hugo Gene Nomenclature Committee (HGNC; http://www.genenames.org/); Online Mendelian Inheritance in Man (OMIN; http://omim.org/); UniProt (http://www.uniprot.org/); National Center for Biotechnology Information (NCBI; http://www.ncbi.nlm.nih.gov/guide/); and PubMed-NCBI (http://www.ncbi.nlm.nih.gov/pubmed)

Table 4.3 Examples of structural hereditary chaperonopathies due to mutations in the genes encoding Groups I and II chaperonins

Gene/protein affected	Disease/syndrome
Chaperonin group I, Hsp60	
Mitochondrial Hsp60 (Cpn60)	Hereditary spastic paraplegia (SPG13); MitCHAP-60 Disease (Pelizaeus-Merzbacher-like)
Chaperonin group II, CCT subunits	
MKKS, BBS10, and BBS12	McKusick-Kaufman (MKKS), and Bardet-Biedl (BBS)
CCT4, CCT5	Hereditary sensory neuropathy

Source Macario AJL, Grippo TM, Conway de Macario E (2005) Genetic disorders involving molecular-chaperone genes: A perspective. Genet Med 7:3–12; Macario AJL, Conway de Macario E (2005) Sick chaperones, cellular stress and disease. New Eng J Med 353:1489–1501; Macario AJL, Conway de Macario E (2007) Chaperonopathies by defect, excess, or mistake. Ann N Y Acad Sci 1113:178–191; and the following Databases: Hugo Gene Nomenclature Committee (HGNC; http://www.genenames.org/); Online Mendelian Inheritance in Man (OMIN; http://omim.org/); UniProt (http://www.uniprot.org/); National Center for Biotechnology Information (NCBI; http://www.ncbi.nlm.nih.gov/guide/); and PubMed-NCBI (http://www.ncbi.nlm.nih.gov/pubmed)

Table 4.4 Human Hsp60 gene and protein, and mutations

Hsp60 gene and protein

Gene	Name	Cytogenetic location	Structure	Protein
HSPD1 HGNC:5261	Hsp60; HSP60; HSPD1; heat-shock 60-Kd protein 1; Chaperonin, 60-Kd; CPN60; Cpn60; GroEL, *E. coli*, homolog of	*2q33.1*	12 exons (first non-coding) Variants: 2	UniProtKB/Swiss-Prot P10809 573 aa Isoforms: 2

Hsp60 gene mutations

Mutation	Disease (mutation phenotype) name (and synonyms)	Phenotype MIM number	Gene Locus MIM number
Val98Ile (V98I) Gln461Glu (Q461E)	Spastic Paraplegia 13, autosomal dominant; SPG13	605280	118190
Asp29Gly (D29G)	Leukodystrophy, Hypomyelinating, 4; HLD4; Mitochondrial HSP60 Chaperonopathy; MitCHAP-60 Disease	612233	118190

Source Mukherjee K, Conway de Macario E, Macario AJL, Brocchieri L (2010) Chaperonin genes on the rise: new divergent classes and intense duplication in human and other vertebrate genomes. BMC Evolutionary Biology 2010 10:64. doi:10.1186/1471-2148-10-64 http://www.biomedcentral.com/1471-2148/10/64; Cappello F, Macario AJL (2012) "Mitochondrial chaperonin Hsp60: locations, functions and pathology", in Houry, W.A. (ed.), *Protein Homeostasis:* The Biomedical and Life Sciences Collection, Henry Stewart Talks Ltd, London (online at http://hstalks.com/bio). Direct talk access links: http://hstalks.com/lib.php?t=HST142.3142 &c=252; and the following Databases: Hugo Gene Nomenclature Committee (HGNC; http://www.genenames.org/); Online Mendelian Inheritance in Man (OMIN; http://omim.org/); UniProt (http://www.uniprot.org/); National Center for Biotechnology Information (NCBI; http://www.ncbi.nlm.nih.gov/guide/); and PubMed-NCBI (http://www.ncbi.nlm.nih.gov/pubmed)

Table 4.5 Examples of human CCT-associated chaperonopathies: Mutations

Name/HGNC/Gene ID/ UniProtKB/Swiss-Prot/ Accession number	Synonyms	Total aa	Chaperonopathies: Mutations
CCT4/1617/10575/ P50991/NM_006430	Chaperonin-containing TCP1, subunit 4; CCT-delta; CCTD; TCPD; TCP-1 delta; Stimulator of tar RNA-binding proteins; SRB	539	MIM ***605142** Distal hereditary sensory neuropathy (mutilated foot) in rat. Cys450-to-tyr **(C450Y)**
CCT5/1618/22948/ P48643/ NM_012073.3	Chaperonin-containing TCP1, subunit 5; CCT-epsilon; CCTE; TCP1E; TCP1 epsilon; KIAA0098	541	MIM ***610150; #256840** Distal hereditary sensory-motor neuropathy. His147-to-arg **(H147R)**

Source Macario AJL, Grippo TM, Conway de Macario E (2005) Genetic disorders involving molecular-chaperone genes: A perspective. Genet Med 7:3–12; Macario AJL, Conway de Macario E (2005) Sick chaperones, cellular stress and disease. New Eng J Med 353:1489–1501; Macario AJL, Conway de Macario E (2007) Chaperonopathies by defect, excess, or mistake. Ann NY Acad Sci 1113:178–191; and the following Databases: Hugo Gene Nomenclature Committee (HGNC; http://www.genenames.org/); Online Mendelian Inheritance in Man (OMIN; http://omim.org/); UniProt (http://www.uniprot.org/); National Center for Biotechnology Information (NCBI; http://www.ncbi.nlm.nih.gov/guide/); and PubMed-NCBI (http://www.ncbi.nlm.nih.gov/pubmed)

Table 4.6 Human MKKS-BBS-associated chaperonopathies: Mutations

Name/HGNC/Gene ID/ UniProtKB/Swiss-Prot/ Accession number	Synonyms	Total aa	Chaperonopathies: Mutations
MKKS (BBS6)/7108/8195/ Q9NPJ1/NM_018848.2 and NM_170784.1	MKKS GENE; MKS; BBS6 gene; McKusick-Kaufman/Bardet-Biedl syndromes putative chaperonin; Bardet-Biedl syndrome 6 protein	570	MIM:604896 gene; 209900 phenotype; 236700 phenotype; McKusick-Kaufman syndrome; MKKS hydrometrocolpos syndrome; hydrometrocolpos, postaxial polydactyly, and congenital heart malformation; HMCS Kaufman-McKusick syndrome. **Y37C** (604896.0003), **T57A** (604896.0010), and **C499S** (604896.0013): increased MKKS degradation and reduced solubility relative to wildtype MKKS, and the mutant **H84Y** (604896.0001). **R155L, A242S**, and **G345E** mutations: increased MKKS degradation only
BBS10/26291/79738/ Q8TAM1/ NM_024685.3	Bardet-Biedl syndrome 10 protein; C12orf58; BBS10 gene; Chromosome 12 ORF 58; FLJ23560	723	MIM:610148 gene; 209900 phenotype; Bardet-Biedl syndrome (similar but not the same as Laurence-Moon syndrome). (BBS; 209900) a 1-Bp insertion at residue 91 leading to premature termination 4 codons later (**C91fsX95**). **VAL11GLY; ARG34PRO; SER303 FS; SER311ALA**
BBS12/26648/166379/ Q6ZW61/ NM_152618.2	Bardet-Biedl syndrome 12 protein; C4orf24; BBS12 gene; FLJ35630; C4ORF24	710	MIM:610683 Bardet-Biedl syndrome 12. **ALA289PRO; ARG355TER; 3-BP DEL, 335TAG; 2-BP DEL, 1114TT; 2-BP DEL, 1483GA; F372fsX373**

Other manifestations of **MKKS-BBS**: retinal dystrophy, mental retardation, mild obesity, diabetes, small testes and genitalia

Source Macario AJL, Grippo TM, Conway de Macario E (2005) Genetic disorders involving molecular-chaperone genes: A perspective. Genet. Med. 7:3–12; Macario AJL, Conway de Macario E (2005). Sick chaperones, cellular stress and disease. New Eng J Med 353:1489–1501; Macario AJL, Conway de Macario E (2007) Chaperonopathies by defect, excess, or mistake. Ann N Y Acad Sci 1113:178–191; and the following Databases: Hugo Gene Nomenclature Committee (HGNC; http://www.genenames.org/); Online Mendelian Inheritance in Man (OMIN; http://omim.org/); UniProt (http://www.uniprot.org/); National Center for Biotechnology Information (NCBI; http://www.ncbi.nlm.nih.gov/guide/); and PubMed-NCBI (http://www.ncbi.nlm.nih.gov/pubmed)

4.3 Chaperonopathies Associated with Mutation of Genes Belonging to the Hsp40, Hsp70, and Super Heavy Groups

Diseases associated with mutations in the genes encoding proteins within the Hsp40 and Hsp70 families, as well in the gene encoding the very large protein, sacsin (now believed to be a chaperone) have been described in Tables 4.7 and 4.8.

The Hsp40/DnaJ extended family. At least 50 genes-proteins have been classified within the Hsp40/DnaJ family. This large group of human genes-proteins includes at least three main subfamilies, A, B, and C. They all have the J-domain, which is involved in the recruitment of Hsp70 (HSPA) to the Hsp70 chaperoning team and stimulates its ATPase activity. The subfamilies A, B, and C have 4, 14, and 22 members at last count, respectively. In addition, there are at least other 10 proteins encoded in the human genome that also have the J-domain but their overall structures are not as clearly identifiable as belonging to this family in comparison with those of the A, B, and C subfamilies. Abnormal members of this family have been found associated with disease, as shown in Table 4.7

Table 4.7 Examples of structural hereditary chaperonopathies of the Hsp40, Hsp70, and super heavy groups

Gene/protein affected	Disease/syndrome
Hsp40 family	
DNAJB6 (Mrj; mDj4)	Autosomal dominant limb-girdle muscular dystrophy (LGMD) with skeletal muscle vacuoles
HSJ1 (DNAJB2; HSJ1; HSPF3; Dnajb10; MDJ8)	Distal hereditary motor neuropathy
DNAJC5 (CLN4, CLN4B, CSP, DNAJC5A, NCL)	Kufs disease
DNAJC6 (mKIAA0473; auxilin)	Juvenile Parkinsonism
Hsp70 family	
STCH (del223V-226L in ATP-binding domain)	Stomach cancer
mtHsp70 (Mortalin)	Parkinson's
Super heavy	
Sacsin (Motifs: DnaJ; HEPN; Ubiquitin-binding; SRR-Hsp90 ATPase like)	Charlevoix-Saguenay (ARSACS)

Source Macario AJL, Grippo TM, Conway de Macario E (2005) Genetic disorders involving molecular-chaperone genes: A perspective. Genet. Med. 7:3–12; Macario AJL, Conway de Macario E (2005). Sick chaperones, cellular stress and disease. New Eng J Med 353:1489–1501; Macario AJL, Conway de Macario E (2007) Chaperonopathies by defect, excess, or mistake. Ann NY Acad Sci 1113:178–191; Macario AJL, Cappello F, Zummo G, Conway de Macario E (2010) Chaperonopathies of senescence and the scrambling of the interactions between the chaperoning and the immune systems. Ann New York Acad Sci 1197: 85–93; and the following Databases: Hugo Gene Nomenclature Committee (HGNC; http://www.genenames.org/); Online Mendelian Inheritance in Man (OMIN; http://omim.org/); UniProt (http://www.uniprot.org/); National Center for Biotechnology Information (NCBI; http://www.ncbi.nlm.nih.gov/guide/); and PubMed-NCBI (http://www.ncbi.nlm.nih.gov/pubmed)

Table 4.8 Sacsin, the sick molecule in sacsinopathies and its mutations causing disease

SACS gene		aa-sequence domain			
Length (bp)	Chr/Exons/kb/ORF/aa/ MW	Name	Location/ Length (aa)	Putative interaction with:	Mutations in (ARSACS)
13,737	13q.11/1/12.8 kb/ORF 11.5 kb/3,834 aa plus 8 upstream of 1/745 aa Total ORF/13.7 kb/ 4,579 aa/437 kDa	Ubiqutin	12–83/72 22–67/46 NP_055178	UBP system; ABD of typical Hsp70s and STCH	
		DnaJ	4,301–4,360/60 4,322–4,370/49 NP_055178	Hsp70	R4325X
		HEPN	4,451–4,567/117 4,451–4,567/117 NP_055178	Nucleotides (ATP)	N4549D
		SRR	N-terminus ~200 × 3	Hsp90-like ATPase action	D168Y

Sacsin and sacsinopathies. **Gene names**: *SACS, ARSACS, KIAA0730, DKFZp686B15167*. GeneID: 26278; HGNC:10519. **Conserved Domains. Ubiquitin** (the ubiquitin family contains multiple ubiquitin-like proteins: SUMO [smt3 homologue], Nedd8, Elongin B, Rub1, and Parkin; **DnaJ**, a DnaJ-class molecular chaperone with C-terminal Zn finger domain; **HEPN**, Higher Eukaryotes and Prokaryotes Nucleotide-binding domain; and **SRR**, Sacsin repeating region. Abbreviations: Chr, Chromosome; ORF, open reading frame; aa, number of amino acids encoded; MW, molecular weight; UBP, ubiquitin proteasome; ABD, ATP-binding domain; ARSACS, autosomal recessive ataxia of Charlevoix-Saguenay *Source* Yoshihisa Takiyama, MD, PhD, Division of Neurology, Department of Internal Medicine, Jichi Medical University, Tochigi 329-0498, Japan; Takiyama Y (1997) Sacsinopathies: sacsin-related ataxia. Cerebellum, 6: 353-359; and the following Databases: Hugo Gene Nomenclature Committee (HGNC; http://www.genenames.org/); Online Mendelian Inheritance in Man (OMIN; http://omim.org/); UniProt (http://www.uniprot.org/); National Center for Biotechnology Information (NCBI; http://www.ncbi.nlm.nih.gov/guide/); and PubMed-NCBI (http://www.ncbi.nlm.nih.gov/pubmed)

4.3.1 Sacsin and Sacsinopathies

The *SACS* gene, encoding the protein sacsin, was initially described to include only one very long exon of 12.8 kb with an open reading frame (ORF) of 11.5 kb. However, later other eight exons were found upstream from the original big exon, making an ORF of 13.7 kb, Table 4.8. Thus far, at least 28 mutations in this big gene (mostly located in the gigantic exon) have been found in Quebec (Canada) and also in other countries such as Italy, Japan, Spain, Tunisia, and Turkey, that are associated with the chaperonopathy named autosomal recessive spastic ataxia of Charlevoix-Saguenay (ARSACS). It is an early onset neurodegenerative disorder characterized by spastic ataxia, dysarthria, nystagmus, distal muscle wasting, finger and foot deformities, and retinal hypermyelination. The histopathology revealed upper vermis atrophy and loss of Purkinje cells in the cerebellum. As said above, most of the mutations found thus far affect the very large exon, but others were also located upstream of it. ARSACS shows some variation in pathological phenotypes, for instance some cases are not early onset but somewhat delayed, there may be absence of retinal hypermyelination, occurrence of intellectual impairment, and even lack of spasticity. It is therefore necessary to be alert to the existence of these genetic and phenotypic variations in clinical practice, and also take them into consideration when planning clinical investigations on this chaperonopathy. It is likely that other mutations will be discovered along with other diverse clinical and histopathological features.

4.4 Diseases Associated with Mutations in the Genes Encoding Dedicated Chaperones

Chaperones are usually considered ubiquitous and promiscuous because they are present almost everywhere inside the cell and interact with a range of nascent polypeptides and protein substrates in need of assistance by a chaperone. Although this may be true to some extent for some chaperones, current research tends to indicate a narrower range of substrates than previously believed and, for some chaperones, it is thought that they recognize and interact with only a small set of substrates, or even a single substrate. These chaperones with one or few substrates can be considered specialized or dedicated chaperones, which it is safe to predict, will increase in number and variey as research advances in the near future. A list of pathological conditions due to mutations in dedicated chaperones is presented in Table 4.9, parts 1 and 2.

Table 4.9 Structural hereditary chaperonopathies due to mutations in the genes encoding dedicated and specialized chaperones, parts 1 and 2

Gene/protein affected	Disease/syndrome
Part1	
Chaperones and chaperone co-factors for microtubule biogenesis	
Co-factor C (RP2)	X-linked retinitis pigmentosa
Cofactor E	Sanjad-Sakati and Kenny-Caffey
	Progressive motor neuronopathy
SPAST/spastin	Autosomal dominant spastic paraplegia-4 (SPG4)
Mitochondrial chaperones	
protein import	
DNAJC19 (TIM14)	Dilated cardiomyopathy with ataxia (DCMA) syndrome
respiratory chain	
BCS1L	GRACILE and Björnstad
FOXRED1	Infantile onset mitochondrial encephalopathy
innermembrane: proteolysis and ribosome assembly	
SPG7/paraplegin	Autosomal recessive spastic paraplegia 7 (SPG7)
ER chaperones	
SIL1/SIL1 (BAB: BiP-associated protein)	Marinesco-Sjogren
Part 2	
Blood cells glycosylating enzyme T-synthase chaperone	
COSMC; C1GALT1-specific chaperone 1 for T-synthase	Autoimmune Tn antigen
Alpha hemoglobin-stabilizing protein	
(AHSP) ERAF	Beta-Thalassemia
Torsin	
TOR1A (DQ2; DYT1)	Torsion dystonia 1
Hsp47 (collagen synthesis)	
SERPINH1	Recessive osteogenesis imperfect
Co-chaperones	
Bag3	Autosomal dominant childhood muscular dystrophy. Dilated cardiomyopathy in adults

(continued)

Table 4.9   (continued)

Gene/protein affected	Disease/syndrome
Anti-oxidant chaperones	
Cu,Zn-superoxide dismutase (SOD1)	Amyotrophic lateral sclerosis (ALS)
CCS (copper chaperone for superoxide dismutase). Mutation Arg163Trp	Alteration in copper homeostasis (CCS binding to SOD1 impaired)
Histone chaperones	
Asf1 (interaction factor Codanin 1)	Congenital dyserythropoietic anemia type I (CDAI)
Non-coding RNA (ncRNA) chaperone	
LARP7	Primordial dwarfism
Ubiquitin–proteasome system	Several neurodegenerative; Lafora; Angelman
Extracellular chaperones	
Clusterin	Pseudoexfoliation (PEX) syndrome/glaucoma. Recurrent hemolytic uremic

Source Macario AJL, Grippo TM, Conway de Macario E (2005) Genetic disorders involving molecular-chaperone genes: A perspective. Genet. Med. 7:3–12; Macario AJL, Conway de Macario E (2005) Sick chaperones, cellular stress and disease. New Eng J Med 353:1489–1501; Macario AJL, Conway de Macario E (2007) Chaperonopathies by defect, excess, or mistake. Ann N Y Acad Sci 1113:178–191; and the following Databases: Hugo Gene Nomenclature Committee (HGNC; http://www.genenames.org/); Online Mendelian Inheritance in Man (OMIN; http://omim.org/); UniProt (http://www.uniprot.org/); National Center for Biotechnology Information (NCBI; http://www.ncbi.nlm.nih.gov/guide/); and PubMed-NCBI (http://www.ncbi.nlm.nih.gov/pubmed)

Further Reading

Section 4.1

STRUCTURAL HEREDITARY CHAPERONOPATHIES: SMALL-SIZE CHAPERONES

Crystallin Functions

Kannan R, Sreekumar PG, Hinton DR (2012) Novel roles for α-crystallins in retinal function and disease. Prog Retin Eye Res 31:576–604

AlphaB-Crystallin and Myopathies

Selcen D (2011) Myofibrillar myopathies. Neuromuscul Disord 21:161–171

CRYAB (HspB5)

Forrest KM, Al-Sarraj S, Sewry C, Buk S, Veronica Tan S, Pitt M, Durward A, McDougall M, Irving M, Hanna MG, Matthews E, Sarkozy A, Hudson J, Barresi R, Bushby K, Jungbluth H, Wraige E (2010) Infantile onset myofibrillar myopathy due to recessive CRYAB mutations. Neuromuscul Disord 21:37–40

Inagaki N, Hayashi T, Arimura T, Koga Y, Takahashi M, Shibata H, Teraoka K, Chikamori T, Yamashina A, Kimura A (2006) Alpha B-crystallin mutation in dilated cardiomyopathy. Biochem Biophys Res Commun 342:379–386. Comment in: Biochem Biophys Res Commun 2006 Aug 11 346(4):1115–1117

Simon S, Fontaine J-M, Martin JL, Sun X, Hoppe AD, Welsh MJ, Benndorf R, Vicart P (2007) Myopathy-associated B-crystallin mutants: abnormal phosphorylation, intracellular location, and interactions with other small heat shock proteins. J Biol Chem 282:34276–34287

Mutation D109H Causes a Multisystemic Disease with Posterior Polar Cataract, Myofibrillar Myopathy and Cardiomyopathy

Sacconi S, Féasson L, Antoine JC, Pécheux C, Bernard R, Cobo AM, Casarin A, Salviati L, Desnuelle C, Urtizberea A (2012) A novel CRYAB mutation resulting in multisystemic disease. Neuromuscul Disord 22:66–72

HSPB1 K141Q Mutation

Ikeda Y, Abe A, Ishida C, Takahashi K, Hayasaka K, Yamada M (2009) A clinical phenotype of distal hereditary motor neuronopathy type II with a novel HSPB1 mutation. J Neurol Sci 277:9–12

HSPB1 S135F Mutation

Almeida-Souza L, Goethals S, de Winter V, Dierick I, Gallardo R, Van Durme J, Irobi J, Gettemans J, Rousseau F, Schymkowitz J, Timmerman V, Janssens S (2010) Increased monomerization of mutant HSPB1 leads to protein hyperactivity in Charcot-Marie-Tooth neuropathy. J Biol Chem 285:12778–12786

Houlden H, Laura M, Wavrant-De Vrièze F, Blake J, Wood N, Reilly MM (2008) Mutations in the HSP27 (HSPB1) gene cause dominant, recessive, and sporadic distal HMN/CMT type 2. Neurology 71:1660–1668

Ikeda Y, Abe A, Ishida C, Takahashi K, Hayasaka K, Yamada M (2009) A clinical phenotype of distal hereditary motor neuronopathy type II with a novel HSPB1 mutation. J Neurol Sci 277:9–12

HSPB6 P20L Mutation

Nicolaou P, Knöll R, Haghighi K, Fan GC, Dorn GW 2nd, Hasenfuß G, Kranias EG (2008) A human mutation in the anti-apoptotic heat shock protein 20 abrogates its cardioprotective effects. J Biol Chem 283:33465–33471

Gain of Function, i.e., Toxicity by Mutant HspB1 and HspB8

Benndorf R (2010) HspB1 and Hsp8 mutations in neuropathies. In: Stephanie Simon, Andre-Partick Arrigo (eds) Small Stress Proteins and Human Diseases, Sect. 2.5. Nova Science Publishers, Inc., pp 301–324, 2010. ISBN: 978-1-61668-198-2

Sun X, Fontaine J-M, Hoppe AD, Carra S, DeGuzman C, Martin JL, Simon S, Vicart P, Welsh MJ, Landry J, Benndorf R (2010) Abnormal interaction of motor neuropathy-associated mutant HspB8 (Hsp22) forms with the RNA helicase Ddx20 (gemin3). Cell Stress Chaperones 15:567–582

Hsp10

Yang K, Meinhardt A, Zhang B, Grzmil P, Adham IM, Hoyer-Fender S (2012) The small heat shock protein odf1/hspb10 is essential for tight linkage of sperm head to tail and male fertility in mice. Mol Cell Biol 32:216–225

PPI Chaperones: FKB10 and FLB14

Baumann M, Giunta C, Krabichler B, Rüschendorf F, Zoppi N, Colombi M, Bittner RE, Quijano-Roy S, Muntoni F, Cirak S, Schreiber G, Zou Y, Hu Y, Romero NB, Carlier RY, Amberger A, Deutschmann A, Straub V, Rohrbach M, Steinmann B, Rostásy K, Karall D, Bönnemann CG, Zschocke J, Fauth C (2012) Mutations in FKBP14 cause a variant of Ehlers-Danlos syndrome with progressive kyphoscoliosis, myopathy, and hearing loss. Am J Hum Genet 90:201–216

Kelley BP, Malfait F, Bonafe L, Baldridge D, Homan E, Symoens S, Willaert A, Elcioglu N, Van Maldergem L, Verellen-Dumoulin C, Gillerot Y, Napierala D, Krakow D, Beighton P, Superti-Furga A, De Paepe A, Lee B (2011) Mutations in FKBP10 cause recessive osteogenesis imperfecta and Bruck syndrome. J Bone Miner Res 26:666–672

Pyott SM, Schwarze U, Christiansen HE, Pepin MG, Leistritz DF, Dineen R, Harris C, Burton BK, Angle B, Kim K, Sussman MD, Weis M, Eyre DR, Russell DW, McCarthy KJ, Steiner RD, Byers PH (2011) Mutations in PPIB (cyclophilin B) delay type I procollagen chain association and result in perinatal lethal to moderate osteogenesis imperfecta phenotypes. Hum Mol Genet 20:1595–1609

Venturi G, Monti E, Carbonare LD, Corradi M, Gandini A, Valenti MT, Boner A, Antoniazzi F (2012) A novel splicing mutation in FKBP10 causing osteogenesis imperfecta with a possible mineralization defect. Bone 50:343–349

Chahal HS, Stals K, Unterländer M, Balding DJ, Thomas MG, Kumar AV, Besser GM, Atkinson AB, Morrison PJ, Howlett TA, Levy MJ, Orme SM, Akker SA, Abel RL, Grossman AB, Burger J, Ellard S, Korbonits M (2011) AIP mutation in pituitary adenomas in the 18th Century and today. N Engl J Med 364:43–50; Comment by Cazabat L, Bouligand J, Chanson P in N Engl J Med 364:1973–1975, 2011

Section 4.2

CHAPERONIN OF GROUP I: HSP60 (CPN60) AND ASSOCIATED CHAPERONOPATHIES

Hsp60-Hsp10 Locus in Humans

Hansen JJ, Bross P, Westergaard M, Nielsen MN, Eiberg H, Borglum AD, Mogensen J, Kristiansen K, Bolund L, Gregersen N (2003) Genomic structure of the human mitochondrial chaperonin genes: HSP60 and HSP10 are localised head to head on chromosome 2 separated by a bidirectional promoter. Hum Genet 112:71–77

Hsp60 in Neurodegeneration

Bross P, Magnoni R, Bie AS (2012) Molecular chaperone disorders: defective Hsp60 in neurodegeneration. Curr Top Med Chem 12:2491–2503

Spastic Paraplegia (SPG)

Fontaine B, Davoine C-S, Durr A, Paternotte C, Feki I, Weissenbach J, Hazan J, Brice A (2000) A new locus for autosomal dominant pure spastic paraplegia, on chromosome 2q24–q34. Am J Hum Genet 66:702–707

Salinas S, Proukakis C, Crosby A, Warner TT (2008) Hereditary spastic paraplegia: clinical features and pathogenetic mechanisms. Lancet Neurol 7:1127–1138

SPG13 (V72I or Val98Ile) if the Mitochondrial Recognition Sequence is Considered in the Counting of Amino Acids of Hsp60

Hansen JJ, Durr A, Cournu-Rebeix I, Georgopoulos C, Ang D, Davoine CS, Brice A, Fontaine B, Gregersen N, Bross P (2002) Hereditary spastic paraplegia SPG13 is associated with a mutation in the gene encoding the mitochondrial chaperonin Hsp60. Am J Hum Genet 70:1328–1332

SPG (Gln461Glu) and Others

Hansen J, Svenstrup K, Ang D, Nielsen MN, Christensen JH, Gregersen N, Nielsen JE, Georgopoulos C, Bross P (2007) A novel mutation in the HSPD1 gene in a patient with hereditary spastic paraplegia. J Neurol 254:897–900

MitCHAP-60 Disease (D29G)

Magen D, Georgopoulos C, Bross P, Ang D, Segev Y, Goldsher D, Nemirovski A, Shahar E, Ravid S, Luder A, Heno B, Gershoni-Baruch R, Skorecki K, Mandel H (2008) Mitochondrial Hsp60 chaperonopathy causes an autosomal-recessive neurodegenerative disorder linked to brain hypomyelination and leukodystrophy. Am J Hum Genet 83:30–42

CHAPERONIN OF GROUP II: CCT, AND ASSOCIATED CHAPERONOPATHIES

CCT

Bouhouche A, Benomar A, Bouslam N, Chkili T, Yahyaoui M (2006) Mutation in the epsilon subunit of the cytosolic chaperonin-containing t-complex peptide-1 (Cct5) gene causes autosomal recessive mutilating sensory neuropathy with spastic paraplegia. J Med Genet 43:441–443

Bouhouche A, Benomar A, Bouslam N, Ouazzani R, Chkili T, Yahyaoui M (2006) Autosomal recessive mutilating sensory neuropathy with spastic paraplegia maps to chromosome 5p15.31–14.1. Eur J Hum Genet 14:249–252

Hsu SH, Lee MJ, Hsieh SC, Scaravilli F, Hsieh ST (2004) Cutaneous and sympathetic denervation in neonatal rats with a mutation in the delta subunit of the cytosolic chaperonin-containing t-complex peptide-1 gene. Neurobiol Dis 16:335–345

Lee MJ, Stephenson DA, Groves MJ, Sweeney MG, Davis MB, An SF, Houlden H, Salih MA, Timmerman V, de Jonghe P, Auer-Grumbach M, Di Maria E, Scaravilli F, Wood NW, Reilly MM (2003) Hereditary sensory neuropathy is caused by a mutation in the delta subunit of the cytosolic chaperonin-containing t-complex peptide-1 (Cct4) gene. Hum Mol Genet 12:1917–1925

Posokhova E, Song H, Belcastro M, Higgins L, Bigley LR, Michaud NA, Martemyanov KA, Sokolov M (2011) Disruption of the chaperonin containing TCP-1 function affects protein networks essential for Rod outer segment morphogenesis and survival. Mol Cell Proteomics 10(1):M110.000570

Satish L, O'Gorman DB, Johnson S, Raykha C, Gan BS, Wang JH, Kathju S (2013) Increased CCT-eta expression is a marker of latent and active disease and a modulator of fibroblast contractility in Dupuytren's contracture. Cell Stress Chaperones. 2013 Jan 6 (Epub ahead of print)

MKKS and BBS

Billingsley G, Bin J, Fieggen KJ, Duncan JL, Gerth C, Ogata K, Wodak SS, Traboulsi EI, Fishman GA, Paterson A, Chitayat D, Knueppel T, Millán JM, Mitchell GA, Deveault C, Héon E (2010) Mutations in chaperonin-like BBS genes are a major contributor to disease development in a multiethnic Bardet-Biedl syndrome patient population. J Med Genet 47:453–463

Katasanis N, Beales PL, Woods MO, Lewis RA, Green JS, Parfrey PS, Ansley SJ, Davidson WS, Lupski JR (2000) Mutations in MKKS cause obesity, retinal dystrophy and renal malformations associated with Bardett-Biedl syndrome. Nature Genet 1:67–70

Katsanis N, Ansley SJ, Badano JL, Eicher ER, Lewis RA, Hoskins BE, Scambler PJ, Davidson WS, Beales PL, Lupski JR (2001) Triallelic inheritance in Bardet-Biedl syndrome, a Mendelian recessive disorder. Science 293:2256–2259

Loktev AV, Zhang Q, Beck JS, Searby CC, Scheetz TE, Bazan JF, Slusarski DC, Sheffield VC, Jackson PK, Nachury MV (2008) A BBSome subunit links ciliogenesis, microtubule stability, and acetylation. Dev Cell 15:854–865

Seo S, Baye LM, Schulz NP, Beck JS, Zhang Q, Slusarski DC, Sheffield VC (2010) BBS6, BBS10, and BBS12 form a complex with CCT/TRiC family chaperonins and mediate BBSome assembly. Proc Natl Acad Sci USA 107:1488–1493

Slavotinek AM, Stone EM, Mykytyn K, Heckenlively JR, Green JS, Heon E, Musarella MA, Parfrey PS, Sheffield VC, Biesecker LG (2000) Mutations in MKKS cause Bardett-Biedl syndrome. Nature Genet 1:15–16

Stoetzel C, Laurier V, Davis EE, Muller J, Rix S, Badano JL, Leitch CC, Salem N, Chouery E, Corbani S, Jalk N, Vicaire S, Sarda P, Hamel C, Lacombe D, Holder M, Odent S, Holder S, Brooks AS, Elcioglu NH, Silva ED, Rossillion B, Sigaudy S, de Ravel TJ, Lewis RA, Leheup B, Verloes A, Amati-Bonneau P, Mégarbané A, Poch O, Bonneau D, Beales PL, Mandel JL, Katsanis N, Dollfus H (2006) BBS10 encodes a vertebrate-specific chaperonin-like protein and is a major BBS locus. Nat Genet 38:521–524. Erratum in: Nat Genet 2006 Jun; 38(6):727. Da Silva, Eduardo (corrected to Silva, Eduardo D)

Stoetzel C, Muller J, Laurier V, Davis EE, Zaghloul NA, Vicaire S, Jacquelin C, Plewniak F, Leitch CC, Sarda P, Hamel C, de Ravel TJ, Lewis RA, Friederich E, Thibault C, Danse JM, Verloes A, Bonneau D, Katsanis N, Poch O, Mandel JL, Dollfus H (2007) Identification of a novel BBS gene (BBS12) highlights the major role of a vertebrate-specific branch of chaperonin-related proteins in Bardet-Biedl syndrome. Am J Hum Genet 80:1–11

Stone DL, Slavotinek A, Bouffard CG, Banerjee-Basu S, Baxevanis AD, Barr M, Biesecker LG (2000) Mutation of a gene encoding a putative chaperonin causes McKusick-Kaufman syndrome. Nature Genet 1:79–82

Zhang Q, Yu D, Seo S, Stone EM, Sheffield VC (2012) Intrinsic protein–protein interaction mediated and chaperonin assisted sequential assembly of a stable Bardet Biedl syndrome protein complex, the BBSome. J Biol Chem 287:20625–20635

Section 4.3

THE HSP40, HSP70 AND SUPER HEAVY GROUPS AND ASSOCIATED CHAPERONOPATHIES

DNAJB6 and Autosomal Dominant Limb-Girdle Muscular Dystrophy (LGMD) with Skeletal Muscle Vacuoles

Harms MB, Sommerville RB, Allred P, Bell S, Ma D, Cooper P, Lopate G, Pestronk A, Weihl CC, Baloh RH (2012) Exome sequencing reveals DNAJB6 mutations in dominantly inherited myopathy. Ann Neurol 71:416–497

Sarparanta J, Jonson PH, Golzio C, Sandell S, Luque H, Screen M, McDonald K, Stajich JM, Mahjneh I, Vihola A, Raheem O, Penttilä S, Lehtinen S, Huovinen S, Palmio J, Tasca G, Ricci E, Hackman P, Hauser M, Katsanis N, Udd B (2012) Mutations affecting the cytoplasmic functions of the co-chaperone DNAJB6 cause limb-girdle muscular dystrophy. Nat Genet 44:450–455

Sato T, Hayashi YK, Oya Y, Kondo T, Sugie K, Kaneda D, Houzen H, Yabe I, Sasaki H, Noguchi S, Nonaka I, Osawa M, Nishino I (2013) DNAJB6 myopathy in an Asian cohort and cytoplasmic/ nuclear inclusions. Neuromuscul Disord 23:269–276. doi:10.1016/j.nmd.2012.12.010

HSJ1 (DNAJB2) in Distal Hereditary Motor Neuropathy

Blumen SC, Astord S, Robin V, Vignaud L, Toumi N, Cieslik A, Achiron A, Carasso RL, Gurevich M, Braverman I, Blumen N, Munich A, Barkats M, Viollet L (2012) A rare recessive distal hereditary motor neuropathy with HSJ1 chaperone mutation. Ann Neurol 71:509–519

DNAJC5 Mutations Causing Autosomal Dominant Kufs Disease

Velinov M, Dolzhanskaya N, Gonzalez M, Powell E, Konidari I, Hulme W, Staropoli JF, Xin W, Wen GY, Barone R, Coppel SH, Sims K, Brown WT, Züchner S (2012) Mutations in the gene DNAJC5 cause autosomal dominant Kufs disease in a proportion of cases: study of the Parry family and 8 other families. PLoS ONE 7(1):e29729. doi:10.1371/journal.pone.0029729

DNAJC6 Mutations Associated with Juvenile Parkinsonism

Edvardson S, Cinnamon Y, Ta-Shma A, Shaag A, Yim YI, Zenvirt S, Jalas C, Lesage S, Brice A, Taraboulos A, Kaestner KH, Greene LE, Elpeleg O (2012) A deleterious mutation in DNAJC6 encoding the neuronal-specific clathrin-uncoating co-chaperone auxilin, is associated with Juvenile Parkinsonism. PLoS ONE 7(5):e36458

Köroğlu C, Baysal L, Cetinkaya M, Karasoy H, Tolun A (2012) DNAJC6 is responsible for juvenile parkinsonism with phenotypic variability. Parkinsonism Relat Disord S1353–8020(12)00439–7. doi:10.1016/j.parkreldis.2012.11.006

STCH

Yamagata N, Furuno K, Sonoda M, Sugimura H, Yamamoto K (2008) Stomach cancer-derived del223V–226L mutation of the STCH gene causes loss of sensitization to TRAIL-mediated apoptosis. Biochem Biophys Res Commun 376:499–503

Hsp72

Voss MR, Gupta S, Stice JP, Baumgarten G, Lu L, Tristan JM, Knowlton AA (2005) Effect of mutation of amino acids 246–251 (KRKHKK) in HSP72 on protein synthesis and recovery from hypoxic injury. Am J Physiol Heart Circ Physiol 289:H2519–H2525

Sacsin and Sacsinopathies

Anderson JF, Siller E, Barral JM (2010) The sacsin repeating region (SRR): a novel Hsp90-related supra-domain associated with neurodegeneration. J Mol Biol 400:665–674

Anderson JF, Siller E, Barral JM (2011) The neurodegenerative-disease-related protein sacsin is a molecular chaperone. J Mol Biol 411:870–884

Guernsey DL, Dubé MP, Jiang H, Asselin G, Blowers S, Evans S, Ferguson M, Macgillivray C, Matsuoka M, Nightingale M, Rideout A, Delatycki M, Orr A, Ludman M, Dooley J, Riddell C, Samuels ME (2010) Novel mutations in the sacsin gene in ataxia patients from Maritime Canada. J Neurol Sci 288:79–87

Hara K, Onodera O, Endo M, Kondo H, Shiota H, Miki K, Tanimoto N, Kimura T, Nishizawa M (2005) Sacsin-related autosomal recessive ataxia without prominent retinal myelinated fibers in Japan. Mov Disord 20:380–382

Kozlov G, Denisov AY, Girard M, Dicaire MJ, Hamlin J, McPherson PS, Brais B, Gehring K (2011) Structural basis of defects in sacsin HEPN domain responsible for the spastic ataxia ARSACS. J Biol Chem 286:20407–20412

Parfitt DA, Michael GJ, Vermeulen EG, Prodromou NV, Webb TR, Gallo JM, Cheetham ME, Nicoll WS, Blatch GL, Chapple JP (2009) The ataxia protein sacsin is a functional cochaperone that protects against polyglutamine expanded ataxin-1. Hum Mol Genet 18:1556–1565

Romano A, Tessa A, Barca A, Fattori F, Fulvia de Leva M, Terracciano A, Storelli C, Maria Santorelli F, Verri T (2013) Comparative analysis and functional mapping of SACS mutations reveal novel insights into sacsin repeated architecture. Hum Mutat 34:525–537 doi:10.1002/humu.22269

Yamamoto Y, Nakamori M, Konaka K, Nagano S, Shimazaki H, Takiyama Y, Sakoda S (2006) Sacsin-related ataxia caused by the novel nonsense mutation Arg4325X. J Neurol 253: 1372–1373

Section 4.4

CHAPERONOPATHIES AFFECTING DEDICATED CHAPERONES

DNAJC19

Mutations in the mitochondrial import inner-membrane translocase subunit TIM14 (DNAJC19) can cause the dilated cardiomyopathy with ataxia (DCMA) syndrome variously accompanied by other anomalies such as anemia, genital anomalies, and methylglutaconic aciduria

Ojala T, Polinati P, Manninen T, Hiippala A, Rajantie J, Karikoski R, Suomalainen A, Tyni T (2012) New mutation of mitochondrial DNAJC19 causing dilated and noncompaction cardiomyopathy, anemia, ataxia, and male genital anomalies. Pediatr Res 72:432–437

FOXRED1

A homozygous mutation (c.1054C>T; p.R352 W) in FOXRED1 was identified in a child with infantile-onset encephalomyopathy. The mutation caused deficiency of Complex I, i.e., the first and largest enzyme in the respiratory chain residing in the inner mitochondrial membrane. Deficiency of Complex I is the most common mitochondrial disorder in childhood as per current literature. However, the molecular basis of the disorder has not been elucidated in most cases. The finding of the mutation referred to above shows how thinking of the role of chaperones in disease may lead to discovering their actual pathogenic role in many of them

Fassone E, Duncan AJ, Taanman JW, Pagnamenta AT, Sadowski MI, Holand T, Qasim W, Rutland P, Calvo SE, Mootha VK, Bitner-Glindzicz M, Rahman S (2010) FOXRED1, encoding an FAD-dependent oxidoreductase complex-I-specific molecular chaperone, is mutated in infantile-onset mitochondrial encephalopathy. Hum Mol Genet 19:4837–4847

SIL1 and Marinesco-Sjögren Syndrome

Marinesco-Sjögren syndrome (MSS) is characterized by degeneration of the cerebellum with cerebellar ataxia, cataracts, psycho-motor development retardation, and muscle hypotonia and weakness. Mutations in Sil1/BAP, a nucleotide-exchange factor for the Hsp70 ortholog present in mitochondria (also named BiP, see human Hsp70 Table), have been implicated in MSS. These mutations result in deletion of most of the Sil1/BAP or affect only five or six amino acids at its C-terminus. All mutations cause a multi-system disorder characteristic of MSS

Anttonen AK, Mahjneh I, Hämäläinen RH, Lagier-Tourenne C, Kopra O, Waris L, Anttonen M, Joensuu T, Kalimo H, Paetau A, Tranebjaerg L, Chaigne D, Koenig M, Eeg-Olofsson O, Udd B, Somer M, Somer H, Lehesjoki AE (2005) The gene disrupted in Marinesco-Sjögren syndrome encodes SIL1, an HSPA5 cochaperone. Nat Genet 37:1309–1311. Comment in: Nat Genet 2005 Dec; 37(12):1302–3

Howes J, Shimizu Y, Feige MJ, Hendershot LM (2012) C-terminal mutations destabilize Sil1/BAP can cause Marinesco-Sjogren Syndrome. J Biol Chem 287:8552–8560

PPIB (cyclophilin B)
Brittle-Bone Disease or Osteogenesis Imperfecta

This condition is characterized by fragile bones with repeated fractures during childhood. Mutations in the PPIB (cyclophilin B) gene have been found associated with abnormalities in type I procollagen chain assemblage resulting in osteogenesis imperfect of various degrees of severity from moderate to lethal

Pyott SM, Schwarze U, Christiansen HE, Pepin MG, Leistritz DF, Dineen R, Harris C, Burton BK, Angle B, Kim K, Sussman MD, Weis M, Eyre DR, Russell DW, McCarthy KJ, Steiner RD, Byers PH (2011) Mutations in PPIB (cyclophilin B) delay type I procollagen chain association and result in perinatal lethal to moderate osteogenesis imperfecta phenotypes. Hum Mol Genet 20:1595–1609

BCSL1

Blázquez A, Gil-Borlado MC, Morán M, Verdú A, Cazorla-Calleja MR, Martín MA, Arenas J, Ugalde C (2009) Infantile mitochondrial encephalomyopathy with unusual phenotype caused by a novel BCS1L mutation in an isolated complex III-deficient patient. Neuromuscul Disord 19:143–146

de Lonlay P, Valnot I, Barrientos A, Gorbatyuk M, Tzagoloff A, Taanman JW, Benayoun E, Chrétien D, Kadhom N, Lombès A, de Baulny HO, Niaudet P, Munnich A, Rustin P, Rötig A (2001) A mutant mitochondrial respiratory chain assembly protein causes complex III deficiency in patients with tubulopathy, encephalopathy and liver failure. Nat Genet 29:57–60

Fernandez-Vizarra E, Bugiani M, Goffrini P, Carrara F, Farina L, Procopio E, Donati A, Uziel G, Ferrero I, Zeviani M (2007) Impaired complex III assembly associated with BCS1L gene mutations in isolated mitochondrial encephalopathy. Hum Mol Genet 16:1241–1252

Hinson JT, Fantin VR, Schönberger J, Breivik N, Siem G, McDonough B, Sharma P, Keogh I, Godinho R, Santos F, Esparza A, Nicolau Y, Selvaag E, Cohen BH, Hoppel CL, Tranebjaerg L, Eavey RD, Seidman JG, Seidman CE (2007) Missense mutations in the BCS1L gene as a cause of the Björnstad syndrome. N Engl J Med 356:809–819

Visapää I, Fellman V, Vesa J, Dasvarma A, Hutton JL, Kumar V, Payne GS, Makarow M, Van Coster R, Taylor RW, Turnbull DM, Suomalainen A, Peltonen L (2002) GRACILE syndrome, a lethal metabolic disorder with iron overload, is caused by a point mutation in BCS1L. Am J Hum Genet 71:863–876

Mitochondrial Hsp70 (mtHsp70) and Parkinson's Disease

Burbulla LF, Schelling C, Kato H, Rapaport D, Woitalla D, Schiesling C, Schulte C, Sharma M, Illig T, Bauer P, Jung S, Nordheim A, Schöls L, Riess O, Krüger R (2010) Dissecting the role of the mitochondrial chaperone mortalin in Parkinson's disease: functional impact of disease-related variants on mitochondrial homeostasis. Hum Mol Genet 19:4437–4452

De Mena L, Coto E, Sánchez-Ferrero E, Ribacoba R, Guisasola LM, Salvador C, Blázquez M, Alvarez V (2009) Mutational screening of the mortalin gene (HSPA9) in Parkinson's disease. J Neural Transm 116:1289–1293

Mitochondrial Chaperone Cofactors

Diaz F, Kotarsky H, Fellman V, Moraes CT (2011) Mitochondrial disorders caused by mutations in respiratory chain assembly factors. Semin Fetal Neonatal Med 16:197–204

Schaefer MK, Schmalbruch H, Buhler E, Lopez C, Martin N, Guénet JL, Haase G (2007) Progressive motor neuronopathy: a critical role of the tubulin chaperone TBCE in axonal tubulin routing from the Golgi apparatus. J Neurosci 27:8779–8789

Tubulin Chaperones

Bartolini F, Tian G, Piehl M, Cassimeris L, Lewis SA, Cowan NJ (2005) Identification of a novel tubulin-destabilizing protein related to the chaperone cofactor E. J Cell Sci 118:1197–1207

Bommel H, Xie G, Rossoll W, Wiese S, Jablonka S, Boehm T, Sendtner M (2002) Missense mutation in the tubulin-specific chaperone E (Tbce) gene in the mouse mutant progressive motor neuronopathy, a model of human motoneuron disease. J Cell Biol 159:563–569

Carranza G, Castaño R, Fanarraga ML, Villegas JC, Gonçalves J, Soares H, Avila J, Marenchino M, Campos-Olivas R, Montoya G, Zabala JC (2012) Autoinhibition of TBCB regulates EB1-mediated microtubule dynamics. Cell Mol Life Sci PMID: 22940919

Courtens W, Wuyts W, Poot M, Szuhai K, Wauters J, Reyniers E, Eleveld M, Diaz G, Nöthen MM, Parvari R (2006) Hypoparathyroidism-retardation-dysmorphism syndrome in a girl: A new variant not caused by a TBCE mutation–clinical report and review. Am J Med Genet A 140:611–617. Comment in: Am J Med Genet A 2007 Feb 1; 143(3):301–302; author reply 303–304

Kortazar D, Fanarraga ML, Carranza G, Bellido J, Villegas JC, Avila J, Zabala JC (2007) Role of cofactors B (TBCB) and E (TBCE) in tubulin heterodimer dissociation. Exp Cell Res 313:425–436

Parvari R, Diaz GA, Hershkovitz E (2007) Parathyroid development and the role of tubulin chaperone E. Horm Res 67:12–21

SPG4 (spastin)

Brugman F, Wokke JH, Scheffer H, Versteeg MH, Sistermans EA, van den Berg LH (2005) Spastin mutations in sporadic adult-onset upper motor neuron syndromes. Ann Neurol 58:865–869

Lim JS, Sung JJ, Hong YH, Park SS, Park KS, Cha JI, Lee JY, Lee KW (2010) A novel splicing mutation (c.870 + 3A>G) in SPG4 in a Korean family with hereditary spastic paraplegia. J Neurol Sci 290:186–189

Magariello A, Muglia M, Patitucci A, Ungaro C, Mazzei R, Gabriele AL, Sprovieri T, Citrigno L, Conforti FL, Liguori M, Gambardella A, Bono F, Piccoli T, Patti F, Zappia M, Mancuso M, Iemolo F, Quattrone A (2010) Mutation analysis of the SPG4 gene in Italian patients with pure and complicated forms of spastic paraplegia. J Neurol Sci 288:96–100

Salinas S, Carazo-Salas RE, Proukakis C, Schiavo G, Warner TT (2007) Spastin and microtubules: Functions in health and disease. J Neurosci Res 85:2778–2782

SPG7 (paraplegin)

Brugman F, Scheffer H, Wokke JH, Nillesen WM, de Visser M, Aronica E, Veldink JH, van den Berg LH (2008) Paraplegin mutations in sporadic adult-onset upper motor neuron syndromes. Neurology 71:1500–1505. Comment in: Neurology 2008 Nov 4; 71(19):1468–1469

McDermott CJ, Dayaratne RK, Tomkins J, Lusher ME, Lindsey JC, Johnson MA, Casari G, Turnbull DM, Bushby K, Shaw PJ (2001) Paraplegin gene analysis in hereditary spastic paraparesis (HSP) pedigrees in northeast England. Neurology 56:467–471

SERPINH1

SERPINH1: This gene, also named Hsp47, heat-shock inducible, encodes various members of the serpin superfamily of serine proteinase inhibitors. The protein is present in the ER lumen, interacts with collagen and, therefore, it is considered a chaperone participating in the maturation of collagen molecules. Patients with rheumatoid arthritis have been found that have anti-serpin autoantibodies but the role of these antibodies in pathogenesis remains to be elucidated

Christiansen HE, Schwarze U, Pyott SM, AlSwaid A, Al Balwi M, Alrasheed S, Pepin MG, Weis MA, Eyre DR, Byers PH (2010) Homozygosity for a missense mutation in SERPINH1, which

encodes the collagen chaperone protein HSP47, results in severe recessive osteogenesis imperfecta. Am J Hum Genet 86:389–398
Makareeva E, Aviles NA, Leikin S (2011) Chaperoning osteogenesis: new protein-folding disease paradigms. Trends Cell Biol 21:168–176

Cosmc

Ju T, Cummings RD (2002) A unique molecular chaperone Cosmc required for activity of the mammalian core 1 beta 3-galactosyltransferase. Proc Natl Acad Sci USA 99:16613–16618

AHSP

A mutation has been studied that alters interaction of AHSP with Hb
Brillet T, Baudin-Creuza V, Vasseur C, Domingues-Hamdi E, Kiger L, Wajcman H, Pissard S, Marden MC (2010) Alpha-hemoglobin stabilizing protein (AHSP), a kinetic scheme of the action of a human mutant, AHSPV56G. J Biol Chem 285:17986–17992

UBIQUITIN-PROTEASOME SYSTEM
Lafora disease

This condition is characterized by progressive myoclonic epilepsy with intracellular polyglucosan inclusions, the Lafora bodies, in various tissues including brain, liver and skin. The disease is caused by mutations in the EPM2A gene, encoding the protein phosphatase, laforin, or in the EPM2B gene, encoding the ubiquitin ligase, malin
Dindot SV, Antalffy BA, Bhattacharjee MD, Beaudet AL (2008) The Angelman syndrome ubiquitin ligase localizes to the synapse and nucleus, and maternal deficiency results in abnormal dendritic spine morphology. Hum Mol Genet 17:111–118
Rao SN, Maity R, Sharma J, Dey P, Shankar SK, Satishchandra P, Jana NR (2010) Sequestration of chaperones and proteasome into Lafora bodies and proteasomal dysfunction induced by Lafora disease-associated mutations of malin. Hum Mol Genet 19:4726–4734

CLUSTERIN (CLU) (see also Sect. 2.1 and Table 7.2)
Clusterin Deficiency

Deficiency in this gene's product has been found associated with pseudoexfoliation syndrome in certain types of glaucoma, and mutations of the gene have been implicated in recurrent hemolytic uremic syndrome
Zenkel M, Kruse FE, Jünemann AG, Naumann GO, Schlötzer-Schrehardt U (2006) Clusterin deficiency in eyes with pseudoexfoliation syndrome may be implicated in the aggregation and deposition of pseudoexfoliative material. Invest Ophthalmol Vis Sci 47:1982–1990

Clusterin Mutation

Ståhl AL, Kristoffersson A, Olin AI, Olsson ML, Roodhooft AM, Proesmans W, Karpman D (2009) A novel mutation in the complement regulator clusterin in recurrent hemolytic uremic syndrome. Mol Immunol 46:2236–2243

Clusterin in Familial Amyloidotic Polyneuropathy (FAP)

Clusterin modulates transthyretin (TTR) aggregate formation and, thus, has a protective role in FAB
Magalhães J, Saraiva MJ (2012) The heat shock response in FAP: the role of the extracellular chaperone clusterin. Amyloid 19:167–170

Clusterin Favors Prostate Cancer Dissemination

Shiota M, Zardan A, Takeuchi A, Kumano M, Beraldi E, Naito S, Zoubeidi A, Gleave ME (2012)
Clusterin mediates TGF-β-induced epithelial-mesenchymal transition and metastasis via
Twist1 in prostate cancer cells. Cancer Res 72:5261–5272

BAG3

Some cases of familial dilated cardiomyopathies have been found associated with mutations in
the BAG3 gene

Arimura T, Ishikawa T, Nunoda S, Kawai S, Kimura A (2011) Dilated cardiomyopathy-
associated BAG3 mutations impair Z-disc assembly and enhance sensitivity to apoptosis in
cardiomyocytes. Hum Mutat 32:1481–1491

McCollum AK, Casagrande G, Kohn EC (2010) Caught in the middle: the role of Bag3 in
disease. Biochem J 425(1):e1–3. Comment on: Biochem J. 2010 Jan 1; 425(1):245–255

Selcen D, Muntoni F, Burton BK, Pegoraro E, Sewry C, Bite AV, Engel AG (2009) Mutation in
BAG3 causes severe dominant childhood muscular dystrophy. Ann Neurol 65:83–89

Tn Syndrome

Ju T, Cummings RD (2005) Protein glycosylation: chaperone mutation in Tn syndrome. Nature
437:1252

Ju T, Aryal RP, Kudelka MR, Wang Y, Cummings RD (2013) The cosmc connection to the tn
antigen in cancer. Dis Markers 2013 Mar 11 [Epub ahead of print]

Codanin 1 (Interaction Factor for the Histone Chaperone Named Anti-Silencing function 1 (Asf1)

Ask K, Jasencakova Z, Menard P, Feng Y, Almouzni G, Groth A (2012) Codanin-1, mutated in
the anaemic disease CDAI, regulates Asf1 function in S-phase histone supply. EMBO J
31:2013–2023. Erratum in: EMBO J 31:3229

SOD1 Mutations in ALS

Furukawa Y, O'Halloran TV (2006) Posttranslational modifications in Cu, Zn-superoxide
dismutase and mutations associated with amyotrophic lateral sclerosis. Antioxid Redox Signal
8:847–867

Copper Chaperone for Superoxide Dismutase (CCS) Mutation Impairs Binding of CCS to SOD1 that may Cause Alterations in Copper Hpmeostasis

Huppke P, Brendel C, Korenke GC, Marquardt I, Donsante A, Yi L, Hicks JD, Steinbach PJ,
Wilson C, Elpeleg O, Møller LB, Christodoulou J, Kaler SG, Gärtner J (2012) Molecular and
biochemical characterization of a unique mutation in CCS, the human copper chaperone to
superoxide dismutase. Hum Mutat 33:1207–1215

LARP7 Mutation and Dwarfism

Alazami AM, Al-Owain M, Alzahrani F, Shuaib T, Al-Shamrani H, Al-Falki YH, Alsheddi T,
Colak D, Alkuraya FS (2012) Loss of function mutation in LARP7, chaperone of 7SK
ncRNA, causes a syndrome of facial dysmorphism, intellectual disability and primordial
dwarfism. Hum Mutat 33:1429–1434

Chapter 5
Other Genetic Chaperonopathies

Abstract In this chapter are presented chaperonopathies in which a genetic mechanism is involved but are different from those discussed in chapter 4. Thus, chaperonopathies due to gene dysregulation such as those observed in aged individuals and in some cases with neurodegenerative diseases (e.g., Alzheimer's, Huntington's, Parkinson's, and other conditions), are presented. Likewise, examples of the impact of chaperone-gene polymorphisms on health and disease are given. The quantitative chaperonopathies attributable to gene dysregulation are discussed.

Keywords Gene dysregulation · Quantitative chaperonopathies · Neurodegenerative diseases · Alzheimer's · Huntington's · Parkinson's · Age-associated diseases · Beta-thalassemia · Genetic polymorphism, longevity · Genetic polymorphism, cancer risk

5.1 Chaperonopathies Due to Gene Dysregulation

Some chaperones may be increased or decreased in certain affected cells and tissues, indicating that there is a quantitative abnormality. This variation in the concentration and total quantity of a chaperone can be due to dysregulation of its gene, which is either over- or down-regulated. Other factors that control protein levels may also be involved, for example, increased or decreased degradation of pertinent mRNA or of the chaperone molecule itself.

Probably most chaperonopathies due to gene dysregulation are genetic, but they can also be acquired (and secondary) if the regulatory proteins (for example HSFs) are post-translationally altered (oxidation, glycation, etc.). Examples of these quantitative chaperonopathies are given in Table 5.1. In addition to the examples listed in this Table, chaperones increase or decrease in a variety of cancers as it will be discussed later (Sect. 7.2). For instance, in cancers of colon, uterine cervix, and prostate, Hsp60 increases gradually as the carcinogenic transformation progresses from normal tissue to hyperplasia, dysplasia, and fully developed

A. J. L. Macario et al., *The Chaperonopathies*, SpringerBriefs in Biochemistry and Molecular Biology, DOI: 10.1007/978-94-007-4667-1_5, © The Author(s) 2013

Table 5.1 Examples of chaperonopathies due to gene dysregulation

Chaperone gene	Abnormal manifestation/disease
hsp70	Constitutive and stress-induced levels: low in the aged
DnaJB6, hsp60; hsp70RY; hsc70; hsp70; hsp60; grp75; grp78; grp94; alpha-B-crystallin	Constitutive and stress-induced expression: decreased or increased (e.g., Alzheimer's; Huntington's; Parkinson's, Reducing Body Myopathy, other neurodegenerative diseases)
hsp70	Stress-induced expression: decreased in cardiopathy of the aged
Alpha hemoglobin-stabilizing protein (AHSP) ERAF	Beta-Thalassemia

Source Macario AJL, Grippo TM, Conway de Macario E (2005) Genetic disorders involving molecular-chaperone genes: A perspective. Genet Med 7:3–12; Macario AJL, Conway de Macario E (2005). Sick chaperones, cellular stress and disease. New Eng J Med 353:1489–1501; and Macario AJL, Conway de Macario E (2007) Chaperonopathies by defect, excess, or mistake. Ann N Y Acad Sci 1113:178–191

carcinoma. It appears that the chaperonin is upregulated and favors tumor growth and dissemination. This is the reason why these tumors are considered possible examples of chaperonopathies by mistake, as explained in Sect. 7.2. Contrarisewise, in other tumors such as cancers of the lung, urinary bladder and oral cavity, Hsp60 decreases as the carcinogenic steps go from normal tissue to carcinoma. In this situation, one may speculate that the chaperonin is involved in anti-cancer mechanisms but is shut off by the malignant machinery. This would be another example of chaperonopathy by gene dysregulation, like those mentioned in the preceding lines, but the mechanism would be different, involving gene downregulation rather than upregulation. These cases, such as lung cancer, in which the levels of one or more chaperones are lower than normal must not be confused with those in which the chaperones might become trapped in protein precipitates, thus, "disappearing" from the cytosol, for example. This capture of chaperone molecules by protein precipitates are probably the mechanism by which some chaperones decrease in a variety of pathologies, such as some neurodegenerative diseases at certain stages of their development.

Still another interesting example of chaperonopathy due to gene dysregulation is provided by the Alpha hemoglobin-stabilizing protein (AHSP), also named Erythroid-associated factor (ERAF) and Erythroid differentiation-related factor (EDRF), whose gene map on chromosome 16 (locus 16p11.2; OMIM 605821). AHSP is an erythroid-specific protein that forms a stable complex with free alpha-hemoglobin but not with beta-hemoglobin or hemoglobin A (alpha2-beta2). In addition, AHSP specifically protects free alpha-hemoglobin from precipitation in solution and in cells. AHSP-gene dosage may modulate pathologic states of alpha-hemoglobin excess such as beta-thalassemia. This is suggested by the fact that although beta-thalassemia is considered to be a monogenic disease, there is diversity of clinical forms among patients who inherit identical mutations in the beta-globin gene. This clinical variability among patients carrying the same mutation clearly indicates that a variety of other determinants, most likely genetic, contribute to the final outcome in terms of phenotype. For instance, it has been proposed that alleles altering the levels or function of AHSP might be the basis for at least some of the clinical variability observed in beta-thalassemia patients. A search for mutations in the AHSP gene did not show any with rare exceptions (see Table 4.9, part 2). Therefore, one can hypothesize that, as a rule, the alteration in the levels of AHSP are due to either regulation of its gene by another, which might be mutated or otherwise altered, or to a post-transcriptional/post-translational mechanism that modulates the levels of AHSP. In any case, reduction in AHSP can be associated with pathology, as shown by research in mice and humans. Hence, conditions such as some forms of thalassemia that are directly associated with AHSP being scarce can be considered chaperonopathies by dysregulation.

There are situations in which alpha-Hb variants are defective in their ability to bind AHSP, causing beta-thalassemia syndromes. In these cases the AHSP is normal in structure and quantity but fails to interact with its substrate, the alpha-hemoglobin chain, because a mutation in it impedes its interaction with AHSP, which results in a chaperonopathy of the type "substrate failure" (see Sect. 5.3).

5.2 Human Chaperone-Gene Polymorphisms

Many studies have shown a positive correlation between a given polymorphism in the coding or regulatory region of a chaperone gene and life span, risk of certain diseases, etc. A few examples are presented in the Table 5.2, and others are mentioned in Further Reading.

Table 5.2 Examples of human chaperone-gene polymorphisms

Genetic polymorphism	Abnormality/disease
hsp70-1 promoter region, allele (A)-110	Does not favor longevity in women
hsp70-Hom, T247C, methionine/493/ threonine	Does not favor longevity
hsp70-1 promoter region, allele I-110	Associates with Parkinson's disease in Taiwanese
hsp70-1 5'UTR region, +190CC genotype and haplotypes including +190C	Associates with higher risk of coronary heart disease vs. +190GG and +190G
CRYAB (HspB5) C-802G G allele (CG/ GG at CRYAB C-802G)	Associates with higher risk of oral cancer

Source Macario AJL, Grippo TM, Conway de Macario E (2005) Genetic disorders involving molecular-chaperone genes: A perspective. Genet. Med. 7:3–12; Macario AJL, Conway de Macario E (2005). Sick chaperones, cellular stress and disease. New Eng J Med 353:1489–1501; Macario AJL, Conway de Macario E (2007) Chaperonopathies by defect, excess, or mistake. Ann N Y Acad Sci 1113:178–191; and the Databse dbSNP (http://www.ncbi.nlm.nih.gov/snp/)

5.3 Chaperonopathies Due to Substrate Mutations that Interfere with the Chaperone-Substrate Interaction

There are cases in which a given chaperone fails to function to no fault of its own but to that of the substrate, which is altered (e.g., mutation) in a way that prevents the chaperone from interacting with it. Possible examples of this type of chaperonopathy by "substrate failure" are some types of beta-thalassemias, defects in tubulin assembly and cataracts, as well as cystic fibrosis and Von Hippel-Lindau disease. These in fact are proteinopathies that overlap to some extent with chaperonopathies in as much as the chaperone cannot exercise its function (for AHSP, see also Sects. 4.4 and 5.1).

Further Reading

Section 5.1

CHAPERONOPATHIES DUE TO GENE DYSREGULATION
DnaJB
This gene is upregulated in Parkinsons disease. DNAJB6 is increased in astrocytes in Parkinson's disease, and it is present in Lewy bodies. Durrenberger PF, Filiou MD, Moran LB, Michael GJ, Novoselov S, Cheetham ME, Clark P, Pearce RK, Graeber MB (2009). DnaJB6 is present in the core of Lewy bodies and is highly up-regulated in parkinsonian astrocytes. J Neurosci Res 87:238–245

Grp78 (HSPA5)
This ER chaperone is increased in hereditary reducing-body myopathy (RBM). Liewluck T, Hayashi YK, Ohsawa M, Kurokawa R, Fujita M, Noguchi S, Nonaka I, Nishino I (2007) Unfolded protein response and aggresome formation in hereditary reducing-body myopathy. Muscle Nerve 35:322–326

AHSP (See also Further Reading for Sects. 2.1, 4.4 and 5.3)
dos Santos CO, Zhou S, Secolin R, Wang X, Cunha AF, Higgs DR, Kwiatkowsk JL, Thein SL, Gallagher PG, Costa FF, Weiss MJ (2008) Population analysis of the alpha hemoglobin stabilizing protein (AHSP) gene identifies sequence variants that alter expression and function. Am J Hematol 83:103–108
Turbpaiboon C, Wilairat P (2010) Alpha-hemoglobin stabilizing protein: molecular function and clinical correlation. Front Biosci 15:1–11
Vasseur-Godbillon C, Marden MC, Giordano P, Wajcman H, Baudin-Creuza V (2006) Impaired binding of AHSP to alpha chain variants: Hb Groene Hart illustrates a mechanism leading to unstable hemoglobins with alpha thalassemic like syndrome. Blood Cells Mol Dis, 37:173–179

AlphaB-Crystallin (CRYAB; HspB5)

This small Hsp is overexpressed in some neurodegenerative disorders. This increase in HspB5 may be paralleled in familial amyloidotic polyneuropathy, which is characterized by deposition of mutated transthyretin in the extracellular space, by activation of the heat-shock factor HSF-1 and increase in Hsp27 and Hsp70. Magalhães J, Santos SD, Saraiva MJ (2010) AlphaB-crystallin (HspB5) in familial amyloidotic polyneuropathy. Int J Exp Pathol 91:515–521

Nopp140

Renvoisé B, Colasse S, Burlet P, Viollet L, Meier UT, Lefebvre S (2009) The loss of the snoRNP chaperone Nopp140 from Cajal bodies of patient fibroblasts correlates with the severity of spinal muscular atrophy. Hum Mol Genet 18:1181–1189

Chaperones in Brain Deficiency

Yoo BC, Fountoulakis M, Dierssen M, Lubec G (2001) Expression patterns of chaperone proteins in cerebral cortex of the fetus with Down syndrome: dysregulation of T-complex protein 1. J Neural Transm. Supplementum 61:321–324

Yoo BC, Kim SH, Cairns N, Fountoulakis M, Lubec G (2001) Deranged expression of molecular chaperones in brains of patients with Alzheimer's disease. Biochem Biophys Res Commun 280:249–258

Section 5.2

HUMAN CHAPERONE-GENE POLYMORPHISMS

Polymorphisms of interest can occur in the promoter-regulatory or in the protein-coding regions of the chaperone genes themselves, or of the regulatory genes (e.g., HSF gene)

CRYAB (HspB5)

It was reported that the G allele of CRYAB C-802G correlated with oral cancer risk. Consequently, it was proposed that this polymorphism could be a useful marker to assess and predict oral cancer recurrence and patient survival

Bau DT, Tsai CW, Lin CC, Tsai RY, Tsai MH (2011). Association of alpha B-crystallin genotypes with oral cancer susceptibility, survival, and recurrence in Taiwan. PLoS ONE 6(9):e16374

Hsp70 FAMILY

Chen CM, Wu YR, Hu FJ, Chen YC, Chuang TJ, Cheng YF, Lee-Chen GJ (2008) HSPA5 promoter polymorphisms and risk of Parkinson's disease in Taiwan. Neurosci Lett 435: 219–222

Dato S, Carotenuto L, De Benedictis G (2007) Genes and longevity: A genetic-demographic approach reveals sex- anda ge-specific gene effects not shown by the case-control approach (AP)E and HSP70.1 loci). Biogerontology 8:31–41

Fürnrohr BG, Wach S, Kelly JA, Haslbeck M, Weber CK, Stach, CM, Hueber, AJ, Graef D, Spriewald, BM, Karin Manger K, Herrmann M, Kaufman KM, Frank SG, Goodmon E, James, JA, Schett G, Winkler TH, Harley, JB, Voll RE (2010) Polymorphisms in the Hsp70 gene locus are genetically associated with systemic lupus erythematosus. Ann Rheum Dis 69:1983–1989

Gombos T, Förhécz Z, Pozsonyi Z, Jánoskuti L, Prohászka Z (2008) Interaction of serum 70-kDa heat shock protein levels and HspA1B (+1267) gene polymorphism with disease severity in patients with chronic heart failure. Cell Stress Chaperones 13:199–206

He M, Guo H, Yang X, Zhang X, Zhou L, Cheng L, Zeng H, Hu FB, Tanguay RM, Wu T (2009) Functional SNPs in HSPA1A gene predict risk of coronary heart disease. PLoS ONE 4(3):e4851

Kakiuchi C, Ishiwata M, Nanko S, Kunigi H, Minabe Y, Nakamura K, Mori N, Fujii K, Umekage T, Tochigi M, Kohda K, Sasaki T, Yamada K, Yoshikawa T, Kato T (2006) Funcional polymorphism of HSPA5: Possible association with bipolar disorder. Biochem Biophys Res Commun 336:1136–1143

Karoly E, Fekete A, Banki NF, Szebeni B, Vannay A, Szabo AJ, Tulassay T, Reusz GS (2007) Heat shock protein 72 (HSPA1B) gene polymorphism and toll-like receptor (TLR) 4 mutation are associated with increased risk of urinary tract infection in children. Pediatr Res 61:371–374

Kawaguchi S, Hagiwara A, Suzuki M (2008) Polymorphic analysis of the heat-shock protein 70 gene (HSPA1A) in Ménière's disease. Acta Otolaryngol 128:1173–1177

Klausz G, Molnár T, Nagy F, Gyulai Z, Boda K, Lonovics J, Mándi Y (2005) Polymorphism of the heat-shock protein gene Hsp70-2, but not polymorphisms of the IL-10 and CD14 genes, is associated with the outcome of Crohn's disease. Scand J Gastroenterol 40:1197–1204

Li JX, Tang BP, Sun HP, Feng M, Cheng ZH, Niu WQ (2009) Interacting contribution of the five polymorphisms in three genes of Hsp70 family to essential hypertension in Uygur ethnicity. Cell Stress Chaperones 14:355–362

Nam SY, Kim N, Kim JS, Lim SH, Jung HC, Song IS (2007) Heat shock protein gene 70-2 polymorphism is differentially associated with the clinical phenotypes of ulcerative colitis and Crohn's disease. J Gastroenterol Hepatol 22:1032–1038

Niino M, Kikuchi S, Fukazawa T, Yabe I, Sasaki H, Tashiro K (2001) Heat shock protein 70 gene polymorphism in Japanese patients with multiple sclerosis. Tissue Antigens 58:93–96

Partida-Rodriguez O, Torres J, Flores-Luna L, Camorlinga M, Nieves-Ramìrez M, Lazcano E, Perez-Rodriguez M (2010) Polymorphisms in TNF and HSP-70 show a significant association with gastric cancer and duodenal ulcer. Int J Cancer 126:1861–1868

Singh R, Kolvraa S, Rattan SIS (2007) Genetics of human longevity with emphasis on the relevance of HSP70 as candidate genes. Front Biosci 12:4504–4513

Spagnolo P, Sato H, Marshall SE, Antoniou KM, Ahmad T, Wells AU, Ahad MA, Lightman S, du Bois RM, Welsh KI (2007) Association between heat shock protein 70/Hom genetic polymorphisms and uveitis in patients with sarcoidosis. Invest Ophthalmol Vis Sci 48:3019–3025

Wu YR, Wang CK, Chen CM, Hsu Y, Lin SJ, Lin YY, Fung HC, Chang KH, Lee-Chen GJ (2004) Analysis of heat-shock protein 70 gene polymorphisms and the risk of Parkinson's disease. Hum Genet 114:236–241

Zouiten-Mekki L, Karoui S, Kharrat M, Fekih M, Matri S, Boubaker J, Filali A, Chaabouni H (2007) Crohn's disease and polymorphism of heat shock protein gene HSP70-2 in the Tunisian population. Eur J Gastroenterol Hepatol 19:225–228

OTHER CHAPERONES

Ising M, Depping AM, Siebertz A, Lucae S, Unschuld PG, Kloiber S, Horstmann S, Uhr M, Müller-Myhsok B, Holsboer F (2008) Polymorphisms in the FKBP5 gene region modulate recovery from psychosocial stress in healthy controls. Eur J Neurosci 28:389–398

Jeng JE, Tsai JF, Chuang LY, Ho MS, Lin ZY, Hsieh MY, Chen SC, Chuang WL, Wang LY, Yu ML, Dai CY, Chang JG (2008) Heat shock protein A1B 1267 polymorphism is highly associated with risk and prognosis of hepatocellular carcinoma: a case-control study. Medicine (Baltimore) 87:87–98

Lu J, Hu Z, Wei S, Wang LE, Liu Z, El-Naggar AK, Sturgis EM, Wei Q (2009) A novel functional variant ($-842G > C$) in the PIN1 promoter contributes to decreased risk of squamous cell carcinoma of the head and neck by diminishing the promoter activity. Carcinogenesis 30:1717–1721

Lu J, Yang L, Zhao H, Liu B, Li Y, Wu H, Li Q, Zeng B, Wang Y, Ji W, Zhou Y (2011) The polymorphism and haplotypes of PIN1 gene are associated with the risk of lung cancer in Southern and Eastern Chinese populations. Hum Mutat 32:1299–1308

Maruszak A, Safranow K, Gustaw K, Kijanowska-Haładyna B, Jakubowska K, Olszewska M, Styczyńska M, Berdyński M, Tysarowski A, Chlubek D, Siedlecki J, Barcikowska M, Zekanowski C (2009) PIN1 gene variants in Alzheimer's disease. BMC Med Genet 12;10:115

Segat L, Pontillo A, Annoni G, Trabattoni D, Vergani C, Clerici M, Arosio B, Crovella S (2007) PIN1 promoter polymorphisms are associated with Alzheimer's disease. Neurobiol Ageing 28:69–74

Section 5.3

Chaperonopathies due to Substrate Mutation that Interferes with the Chaperone-Substrate Interaction

Altschuler GM, Dekker C, McCormack EA, Morris EP, Klug DR, Willison KR (2009) A single amino acid residue is responsible for species-specific incompatibility between CCT and alpha-actin. FEBS Lett 583:782–786

Feldman DE, Thulasiraman V, Ferreyra RG, Frydman J (1999) Formation of the VHL-elongin BC tumor suppressor complex is mediated by the chaperonin TRiC. Mol Cell 4:1051–1061

Hutt DM, Martino Roth D, Chalfant M, Youker RT, Matteson J, Brodsky JL, Balch WE (2012) FKBP8 Peptidyl-Prolyl-isomerase activity manages a late stage of CFTR folding and stability. J Biol Chem 287:21914–21925

Moreau KL, King JA (2012) Cataract-causing defect of a mutant γ-Crystallin proceeds through an aggregation pathway which bypasses recognition by the α-Crystallin chaperone. PLoS ONE. 7(5):e37256

Rosser MF, Grove DE, Chen L, Cyr DM (2008) Assembly and misassembly of CFTR: folding defects caused by deletion of F508 occur before and after the calnexin-dependent association of MSD1 and MSD2. Mol Biol Cell 19:4570–4279

Saxena A, Banasavadi-Siddegowda YK, Fan Y, Bhattacharya S, Roy G, Giovannucci DR, Frizzell RA, Wang X (2012) Human heat shock protein 105/110 kDa (Hsp105/110) regulates biogenesis and quality control of misfolded cystic fibrosis transmembrane conductance regulator at multiple levels. J Biol Chem 287:19158–19170

Sims KB (2001) Von Hippel-Lindau disease: gene to bedside. Curr Opin Neurol 14:695–703

Speisky D, Duces A, Bieche I, Rebours V, Hammel P, Sauvanet A, Richard S, Bedossa P, Vidaud M, Murat A, Niccoli P, Scoazec JY, Ruszniewski P, Couvelard A (2012) Molecular profiling of pancreatic neuroendocrine tumors in sporadic and von Hippel-Lindau patients. Clin Cancer Res 18:2838–2849

Tian G, Kong XP, Jaglin XH, Chelly J, Keays D, Cowan NJ (2008) A pachygyria-causing alpha-tubulin mutation results in inefficient cycling with CCT and a deficient interaction with TBCB. Mol Biol Cell 19:1152–1161

Wang Y, Tian G, Cowan NJ, Cabral F (2006) Mutations affecting beta-tubulin folding and degradation. J Biol Chem 281:13628–13635

Chapter 6
Acquired Chaperonopathies

Abstract The process of aging affects many proteins, including molecular chaperones, as discussed in this chapter. Aging is perhaps a good example of multiple acquired chaperonopathies with a combination of aberrant post-translational modifications and gene dysregulation, in which the interaction of the chaperoning and the immune systems is scrambled. This would result in generalized inflammation and other alterations that characterize the functional decline, frailty and sarcopenia, observed as age advances. Here, replacement chaperonotherapy has potential.

Keywords Acquired chaperonopathies · Post-translational modifications · Oxidation · Age-associated diseases · Cataracts · Retinopathy · Inclusion body myositis · Poor chaperoning · Chaperones, low levels · Sick chaperones · Chaperone deficiency · Abnormal Hsp · Stress · Chaperonotherapy · ER-associated degradation (ERAD) · Reactive oxygen species (ROS)

6.1 Chaperonopathies of the Aged

The process of ageing is accompanied by a general decline in many physiological functions in part due to aberrant post-translational modifications of many proteins, including the chaperones. Precisely, when functional chaperones are most needed because of an increasing amount of damaged proteins in need of chaperones, these are also damaged, a state of affairs typical of the aging eye lens and, possibly many other tissues. In this and similar situations characterized by relative scarcity of fully functional, healthy chaperones, replacement chaperonotherapy holds a promising future, Table 6.1.

A. J. L. Macario et al., *The Chaperonopathies*, SpringerBriefs in Biochemistry   71
and Molecular Biology, DOI: 10.1007/978-94-007-4667-1_6, © The Author(s) 2013

Table 6.1 Examples of chaperonopathies of the aged

Gene/protein affected	Disease/syndrome
sHsp (crystallin family)	
Eye lens Beta-crystallin: post-translational modifications (overload)[a]	Cataracts
Eye lens Alpha-crystallin: mutations	Cataracts
Retinal Alpha-A-crystallin: post-translational modifications	Retinopathy
Alpha-B-crystallin	Increased in muscle: inclusion body myositis
Hsp70(DnaK)	Low constitutive levels and poor response to stress
Hsp90	Poor chaperoning ability

[a] Beta-crystallin is not a chaperone but is listed here because it may cause a chaperonopathy indirectly. The increase of abnormal Beta-crystallin molecules with age in the eye lens may reach a point at which all the available Alpha-crystallin molecules (chaperones in this case dedicated to the Beta-crystallin) are no longer enough to deal with the demand. This would be a chaperonopathy with quantitative deficiency of Alpha-crystallin due to substrate excess, i.e., chaperonopathy by substrate overload.

Source Macario AJL, Grippo TM, Conway de Macario E (2005) Genetic disorders involving molecular-chaperone genes: A perspective. Genet Med 7:3–12; Macario AJL, Conway de Macario E (2005) Sick chaperones, cellular stress and disease. New Eng J Med 353:1489–1501; Macario AJL, Conway de Macario E (2007) Chaperonopathies by defect, excess, or mistake. Ann N Y Acad Sci 1113:178–191; Cappello F, Di Stefano A, Conway de Macario E, Macario AJL (2010) Hsp60 and Hsp10 in ageing. In "Heat shock proteins and whole body physiology." Edited by Alexzander A. A. Asea and Bente K. Pedersen, Springer-Verlag, Berlin, Heidelberg, Germany. Vol. 5, Part 3, Chapter 23, pp 401–426, 2010. http://www.springerlink.com/content/q87536k412x74772/; and Csermely P (2001) Chaperone overload is a possible contributor to 'civilization diseases'. Trends Genet 17:701–704

6.2 Acquired Chaperonopathies Associated with Ageing: Possible Pathogenic Mechanism

Ageing, as mentioned earlier, is characterized by alterations in proteins, including the chaperones, which lead to a general decline of function in all tissues and organs. These chaperonopathies are, in principle, amenable to chaperonotherapy, Fig. 6.1.

> **THE AGEING PROCESS OF PROTEINS**
>
> Accumulation of mutations
>
> Oxidation, other
>
> Ageing of chaperones: mutations, oxidation (other) = chaperonopathies
>
> Young chaperones to the rescue
>
> Chaperonotherapy: gene and/or protein

Further Reading

Sections 6.1–6.2

Chaperonopathies Associated with Ageing and Possible Pathogenic Mechanism

Arslan MA, Csermely P, Soti C (2006) Protein homeostasis and molecular chaperones in ageing. Biogerontology 7:383–389

Bhattacharyya J, Shipova EV, Santhoshkumar P, Sharma KK, Ortwerth BJ (2007) Effect of a single AGE modification on the structure and chaperone activity of human alphaB-crystallin. Biochemistry 46:14682–14692

Cloos PA, Christgau S (2004) Post-translational modifications of proteins: implications for ageing, antigen recognition, and autoimmunity. Biogerontology 5:139–158

Fan X, Zhang J, Theves M, Strauch C, Nemet I, Liu X, Qian J, Giblin FJ, Monnier VM (2009) Mechanism of lysine oxidation in human lens crystallins during ageing and in diabetes. J Biol Chem 284:34618–34627

Kamei A, Iwase H, Masuda K (1997) Cleavage of amino acid residue(s) from the N-terminal region of alpha A- and alpha B-crystallins in human crystalline lens during ageing. Biochem Biophys Res Commun 231:373–378

Macario AJL, Conway de Macario E (2001) Molecular chaperones and age-related degenerative disorders. Adv Cell Ageing Gerontol 7:131–162

Nardai G, Csermely P, Soti C (2002) Chaperone function and chaperone overload in the aged. A preliminary analysis. Exp Gerontol 37:1257–1262

Nardai G, Vegh EM, Prohaszka Z, Csermely P (2006) Chaperone-related immune dysfunction: an emergent property of distorted chaperone networks. Trends Immunol 27:74–79

Soti C, Csermely P (2000) Molecular chaperones and the ageing process. Biogerontology 1:225–233

Soti C, Csermely P (2003) Ageing and molecular chaperones. Exp Gerontol 38:1037–1040

Soti C, Csermely P (2007) Ageing cellular networks: chaperones as major participants. Exp Gerontol 42:113–119

Unofolded Protein Response and Ageing

Brown MK, Naidoo N (2012) The endoplasmic reticulum stress response in aging and age-related diseases. Front Physiol 3:263. doi:10.3389/fphys.2012.00263

◀**Fig. 6.1** It is believed that, as age progresses, many proteins suffer post-translational modifications that hinder their functionality. Chaperones, being proteins, also are affected by this age-related damage with impairment of function. For example, failure of chaperones in the endoplasmic reticulum (ER) may lead to accumulation of misfolded proteins in this organelle and, thus, cause an unfolded protein response (UPR), with serious deleterious consequences for the aging organism. Another aspect of ageing is oxidative stress, in which the generation and accumulation of toxic forms of oxygen (reactive oxygen species or ROS) play a pathogenic role. Most likely, the natural anti-oxidants present in the cell to scavenge or inactivate these toxic forms of oxygen are also defective do to senescence. In principle, all these age-related chaperonopathies are amenable to replacement therapy, using chaperone genes or proteins to replenish the stock of functional chaperones. This form of chaperonotherapy is a promising avenue for research in the near future. *Source* Macario AJL, Conway de Macario E (2002) Sick chaperones and ageing: A perspective. Ageing Res Reviews 1:295–311; and Macario AJL, Conway de Macario E (2007) Chaperonopathies and chaperonotherapy. FEBS Lett 581:3681–3688

Hagiwara M, Nagata K (2012) Redox-dependent protein quality control in the ER: folding to degradation. Antioxid Redox Signal 16:1119–1128

Oxidative Stress in Ageing, Anti-Oxidants, and Participation of Chaperones

Andrade LN, Nathanson JL, Yeo GW, Menck CF, Muotri AR (2012) Evidence for premature aging due to oxidative stress in iPSCs from Cockayne syndrome. Hum Mol Genet 21:3825–3834
Brinkmann C, Brixius K (2013) Peroxiredoxins and sports: new insights on the antioxidative defense. J Physiol Sci 63:1–5. doi:10.1007/s12576-012-0237-4
Chen F, Yu Y, Qian J, Wang Y, Cheng B, Dimitropoulou C, Patel V, Chadli A, Rudic RD, Stepp DW, Catravas JD, Fulton DJ (2012) Opposing actions of heat shock protein 90 and 70 regulate nicotinamide adenine dinucleotide phosphate oxidase stability and reactive oxygen species production. Arterioscler Thromb Vasc Biol 32:2989–2999

CHIP Deficiency Accelerates Ageing

Min JN, Whaley RA, Sharpless NE, Lockyer P, Portbury AL, Patterson C (2008) CHIP deficiency decreases longevity, with accelerated ageing phenotypes accompanied by altered protein quality control. Mol Cell Biol 28:4018–4025

Decreased Efficiency of Chaperones with Age

Nuss JE, Choksi KB, Deford JH, Papaconstantinou J (2008) Decreased enzyme activities of chaperones PDI and BiP in aged mouse livers. Biochem Biophys Res Commun 365:355–361

Deamidated Hsp60, i.e., A Chaperone with a Pathological Post-Translational Modification (PTM), Could be One of the Autoantigens Responsible for an Autoimmune Pathogenesis of Rheumatoid Arthritis and Age-Related Pathological Conditions

Goëb V, Thomas-L'otellier M, Daveau R, Charlionet R, Fardellone P, Le Loët X, Tron F, Gilbert D, Vittecoq O (2009) Candidate autoantigens identified by mass spectrometry in early rheumatoid arthritis are chaperones and citrullinated glycolytic enzymes. Arthritis Res Ther 11:R38

Telomerases, Hsp90, and Ageing

DeZwaan DC, Toogun OA, Echtenkamp FJ, Freeman BC (2009) The Hsp82 molecular chaperone promotes a switch between unextendable and extendable telomere states. Nat Struct Mol Biol 16:711–716
Toogun OA, Dezwaan DC, Freeman BC (2008) The hsp90 molecular chaperone modulates multiple telomerase activities. Mol Cell Biol 28:457–467
Woo SH, An S, Lee HC, Jin HO, Seo SK, Yoo DH, Lee KH, Rhee CH, Choi EJ, Hong SI, Park IC (2009) A truncated form of p23 down-regulates telomerase activity via disruption of Hsp90 function. J Biol Chem 284:30871–30880

Chapter 7
Other Types of Chaperonopathies

Abstract A mechanism causing a chaperonopathy that is introduced in this chapter consists of the absence of a chaperone from the place where it is needed (i.e., chaperonopathies by misplacement). Also in this chapter are discussed the unfolded-protein response (UPR), chaperone-mediated autophagy (CMA), and illustrative examples of chaperonopathies by mistake, or collaborationism. In these conditions, one or more chaperones, apparently normal in structure, perform functions that favor disease rather than the contrary, hence the name of chaperonopathy by mistake or collaborationism (a molecule that ought to protect the cell and the organism promotes pathogenesis instead). Many examples of chaperonopathies by mistake have been identified involving various chaperones and co-chaperones, including a variety of cancers, and inflammatory and autoimmune conditions. The participation of Hsp60 in these disorders is analyzed in some detail. The potential role of this chaperone as autoantigen and/or as signal molecule is brought up to central stage in certain cancers, myasthenia gravis, Hashimoto's thyroiditis, chronic obstructive pulmonary disease (COPD), and ulcerative colitis.

Keywords Chaperonopathies by misplacement · Unfolded-protein response (UPR) · Chaperone-mediated autophagy (CMA) · Chaperonopathies by mistake · Chaperones in carcinogenesis · Myelodysplasia · Pelizaeus-Merbacher · Autoantigens · Signal molecules · Human-bacterial Hsp60 crossreaction · Autoimmunity · Inflammation · Vasculitis · Arthritis · Myasthenia gravis · Hashimoto's thyroiditis · Chronic obstructive pulmonary disease · Ulcerative colitis · ER-associated degradation (ERAD) · Apoptosis

7.1 The Consequences of a Chaperone not Being in the Right Place at the Right Time

Some chaperonopathies are characterized not by a structural-function deficiency but by the chaperone not being present at the place in which it has to function when needed (Table 7.1).

Table 7.1 Chaperonopathies by misplacement

Chaperone	Abnormality	Disease/syndrome
Hsp70	Alteration of Hsp70 cytosolic-nuclear shuttling Erythroid cell dysplasia	Myelodysplastic syndromes (MDS)
Mutants of the proteolipid protein 1 (PLP1) gene induce depletion of ER chaperones	ER chaperones insufficient to handle accumulation of misfolded proteins in the organelle: failure of UPR (unfolded protein response)	Pelizaeus-Merbacher disease

Chaperone misplacement. Erythroid cell maturation in human involves the transcription factor GATA-1 and temporary caspase-3 activation. The former is protected in the nucleus from caspase-3 mediated cleavage by interaction with the chaperone Hsp70, which has to translocate from the cytosol. In the erythroid cell dysplasia observed in myelodysplastic syndromes (MDS), Hsp70 fails to move into the nucleus and GATA-1 cleavage occurs, leading to erythroid differentiation impairment

Source Frisan E, Vandekerckhove J, de Thonel A, Pierre-Eugène C, Sternberg A, Arlet JB, Floquet C, Gyan E, Kosmider O, Dreyfus F, Gabet AS, Courtois G, Vyas P, Ribeil JA, Zermati Y, Lacombe C, Mayeux P, Solary E, Garrido C, Hermine O, Fontenay M (2012). Defective nuclear localization of Hsp70 is associated with dyserythropoiesis and GATA-1 cleavage in myelodysplastic syndromes. Blood 119:1532–4220; doi: 10.1182/blood-2011-03-343475; and Numata Y, Morimura T, Nakamura S, Hirano E, Kure S, Goto YI, Inoue K (2013) Depletion of molecular chaperones from the endoplasmic reticulum and fragmentation of the Golgi apparatus associated with pathogenesis in Pelizaeus-Merzbacher disease. J Biol Chem 288:7451–7466. doi:10.1074/jbc.M112.435388

7.2 Quantitative Abnormalities and Chaperonopathies by Mistake or Collaborationism

Some chaperonopathies occur when a normal chaperone functions to favor a pathogenic virus, or a pathologic cell, e.g., a cancer cell (Table 7.2), or a microbial infection and its consequences, such as inflammation (see Sect. 7.3). This chaperonopathies are called by mistake or collaborationism. **Definition of collaborationism.** Collaborationism: act of cooperating traitorously with an enemy that is occupying your country. Wordnet.princeton.edu/perl/webwn. Collaborationism, as a pejorative term, can describe the treason of cooperating with enemy forces occupying one's country. En.wikipedia.org/wiki/Collaborationism.

The possible causes and mechanisms of quantitative changes of chaperones were discussed earlier (see Sect. 3.1).

In the cells of some types of cancer chaperones are not only increased but they also may be post-translationally modified (at least a portion of all the molecules of any given chaperone), and their distribution is also changed in comparison with normal cells. For instance, a mitochondrial chaperone such as Hsp60 (Cpn60) is increased in colon cancer cells and is present outside the mitochondria, in the cytosol and the cell membrane. Similarly, Grp78 (Bip; HSPA5; see Table 2.5), an ER resident, is increased in the cells of ovarian cancer and is expressed also on the surface of these cells. A form of Hsp70 methylated at lysine 561 has been found in the nucleus of cancer cells, although the canonical residence of the normal chaperone, unmethylated at position 561, is the cytosol. Furthermore, tumor cells

Table 7.2 Examples of quantitative abnormalities of chaperones in cancer and of chaperonopathies by mistake or collaborationism

Hsp	Increased in cancer of	Decreased in cancer of	Required for cell growth in cancer of
Hsp10	Large bowel (colon, rectum); uterus (exocervix); prostate	Lung (bronchial epithelium); urinary bladder	
Hsp60	Large bowel (colon, rectum); breast; uterus (exocervix); prostate	Lung (bronchial epithelium); urinary bladder	
Hsp70	Breast; bladder (urothelial carcinoma) (Hsp70-2/HSPA2); CNS (glioma; Hsp70); uterine cervix (Hsp70-2); prostate and colon (Hsp70 and Hsp90; and HOP); ovary (Grp78); prostate and multiple myeloma (Hsp70 and Hsp90)		Breast (Hsp70/HSP70A; Hsp70-2/HSPA2; Hsc70/HSPA8); bladder (Hsp70-2/HSPA2); CNS (glioma); Hsp70-2 (uterine cervix); Hsp70 and Hsp90 stabilize metastasis-promoting protein WASF3 in prostate cancer cells; Hsp72 and Hsp73 maintain Hsp90 and survival of myeloma cells
Hsp90; p23; CDC37	Breast (p23: poor prognosis). Myeloid leukemia (+CDC37); pancreas (endocrine-PET; Hsp90 alpha and beta); myelodysplastic syndrome-acute myeloid leukemia; multiple myeloma (with Hsp70)		CDC37 protects the oncogenic protein FOP2-FGFR 1 in leukemia; Hsp70 and Hsp90 stabilize metastasis-promoting protein WASF3 in prostate cancer cells; Hsp72 and Hsp73 maintain Hsp90 and survival of myeloma cells
BAG3; BAG5	Brain (glioblastomas); prostate		Brain (glioblastomas)
HYOU-1	Breast (invasiveness + MMP2); brain (angiogenesis + VEGF)		
Pin1	Lung: non-small cell lung carcinoma (poor prognosis)		
TRAP1 (Mit-Hsp75)			Colon-rectum (anti-apoptotic, drug resistance)
CCT6A	Colon-rectum		Colon-rectum (in vitro)
Clusterin	Breast (unmethylated promoter); prostate		Favors prostate cancer metastasis

Source Macario AJL, Conway de Macario E (2007) Chaperonopathies by defect, excess, or mistake. Ann New York Acad Sci 1113: 178–191

secrete chaperones, for example, Hsp60 (see Chap. 9). The role of these expatriate chaperones is still poorly understood but most likely is very important and will be the target of much investigation in the immediate future.

The roles of chaperones in cancer are diverse, varying with tumor type and stage and going from favoring tumor growth and dissemination (i.e., chaperonopathies by mistake or collaborationism; see right column in Table 7.2, which may be due to some post-translational modifications of the chaperone molecule) to participating in anti-tumor mechanisms, for instance by promoting tumor-cell apoptosis. One essential mechanism involved is the unfolded protein response (UPR) centered in the ER. Cancer cells, and normal cells around them, are stressed by a variety of intracellular stressors, e.g., accumulation of unfolded polypeptides at a speed above normal due to the fast cancer-cell metabolism, and also stressors present in the extracellular environment that might be released by cancer cells. The UPR engages ER chaperones, such as Grp78, Grp94, PDI, and calreticulin (see Tables 2.1, 2.5, Part 1, and 2.6) along with various kinases. The result is that some chaperones accumulate in the ER, migrate to the plasma-cell membrane, and participate in apoptosis; they may also exit the cell and reach other cells in the vicinity or far away via circulation (see Chap. 9). Important questions that will no doubt provoke the interest of scientists, pathologists, and clinicians and, consequently fuel research projects, pertain to the destinations of these extracellular chaperones after they leave their cells of origin (where do they go?) and to their effects on the cells or structures they meet when they reach the intended targets (what do they do there?).

To understand more precisely the role of chaperones in cancer it is necessary to examine the whole picture of interactions between the chaperoning system, the protein-degradation and elimination machineries, and the apoptotic pathways. While chaperones are typically involved in assisting polypeptides in their folding process and in maintaining a physiological pool of correctly folded proteins, they also participate in processes that lead to the elimination of defective or unwanted protein molecules (see Fig. 2.2). The latter elimination is achieved by various mechanisms involving the ubiquitin–proteasome system (UPS), autophagy, and the unfolded protein response (UPR) (see above; and Sect. 6.2 for a discussion of UPR in ageing, which predisposes to cancer). Autophagy includes microautophagy (direct passage of proteins into lysosomes) and chaperone-mediated autophagy (CMA). In this, chaperones (e.g., Hsc70, also named HSPA8; see Table 2.5, Part 2) selectively bind proteins with the short sequence segment formed by the amino acids lysine, phenylalanine, glutamic acid, arginine, and glutamine, exposed, i.e., accessible to interaction with other molecules. This binding can be followed by internalization of the protein by the lysosome with participation of the LAMP-2 (lysosome-associated membrane protein type 2A) receptor. On the other hand, the UPS may be insufficient if the generation of proteins destined to elimination is increased beyond the capability of this system. If, in addition, the UPR is induced by accumulation of damaged proteins inside the ER, autophagy may come to the rescue and clean the cell of unwanted proteins. However, if autophagy also results insufficient, then the apoptotic pathway is triggered for elimination of the entire cell. Cancer cells that manage to keep all the above processes under control survive and proliferate, ultimately killing the organism in which they reproduce. For an integrated view of the interactions and overlapping of the chaperoning system, including protein degradation and elimination, with the immune system and cancer, see Chap. 8, particularly Fig. 8.1. Obviously, it is necessary to make progress in dissecting the molecular details of the interactions between the chaperoning and the ubiquitin–proteasome systems and the various forms of autophagy and the apoptotic cascade. It will eventually be discovered which specific steps in the whole process can be targeted to induce cancer-cell death by using anticancer agents. Among these, anti-chaperone compounds are worth investigating since, as seen above and throughout this book, chaperones play critical roles at various points in the whole network.

HYOU-1 in cancer invasion. HYOU-1 or Hypoxia up-regulated protein 1 is also named 150 kDa oxygen-regulated protein, ORP150, 170 kDa glucose-regulated protein, GRP170, 170 kDa oxygen-regulated protein precursor, DKFZp686N08236. This molecule is an inducible endoplasmic reticulum (ER) chaperone that is upregulated after various types of stress and has a cytoprotective role in renal, neural, and cardiac models of ischemia–reperfusion injury. A positive correlation between ORP150 and MMP-2 expression suggests that ORP150 acts as a molecular chaperone for MMP-2 secretion and thus tumour invasion. Names for MMP2: Matrix metallopeptidase 2; Gelatinase A; 72 kDa gelatinase; 72 kDa type IV collagenase; 72 kDa type IV collagenase precursor; CLG4; CLG4A; MMP-2; MMP-II, MONA; TBE-1. It was demonstrated in vitro with cell cultures that ORP150 plays a determinant role in tumor-mediated angiogenesis via VEGF processing.

Source Ozawa K, Tsukamoto Y, Hori O, Kitao Y, Yanagi H, Stern DM, Ogawa S (2001) Regulation of tumor angiogenesis by oxygen-regulated protein 150, an inducible endoplasmic reticulum chaperone. Cancer Res 61:4206–4213

7.3 Participation of Chaperones in Autoimmunity, and Inflammation

7.3.1 The Non-Chaperoning Roles of Chaperones: Interaction with the Immune System

The canonical functions of chaperones pertain to protein folding, refolding, translocation, degradation and, according to recent research, selective autophagy (i.e., maintenance of protein homeostasis) but they also have many other functions unrelated to protein homeostasis, for instance interaction with the immune system (Table 7.3).

The interplay between the chaperoning and the immune systems of any given organism can in principle occur at least in three main forms. One form involves adaptive immunity (see Sects. 7.3 and 7.4). A chaperone acts as immunogen, undergoes antigen processing and presentation and antibodies are generated

Table 7.3 Examples of reported Hsp60-immune system interactions: Target cells, effects, and receptors

Cell	Effect	Receptor
Macrophages and dendritic cells	IFN-α production that leads to antigen-dependent T cell release of IFN-gamma	Unknown (TLR4 independent)
Macrophages	TNF-α production	TLR4
Neutrophilic granulocytes	Production of ROS and release of primary-granule enzymes	Unknown
Peripheral blood mononuclear cells	TNF-α secretion	Unknown
T cells	Tregs population enhancement	TLR2 and/or TCR
T cells	Proliferation	CD54 RA + RO-
T cells	Down-regulation of chemokines receptor expression (CXCR4 and CCR7)	TLR2
T cells	Production of IFNγ	Unknown
CD4 + T cells	IL-10 secretion	CD30
Microglia	Increase IL-1β, TNF-α, NO, and ROS	LOX-1
Adipocytes	Adipocyte inflammation; muscle-cell insulin resistance	Unknown (interacts with Hsp60 N-terminus)
Cardiomyocytes; macrophages	IRAK-1 activation	TLR4
Endothelial cells, smooth muscle cells, dendritic cells	Cross-presentation on MHC class I molecules	LOX-1

Source Macario AJL, Conway de Macario E (2005) Sick chaperones, cellular stress and disease. New Eng J Med 353: 1489–1501; Macario AJL, Conway de Macario E (2007) Chaperonopathies by defect, excess, or mistake. Ann New York Acad Sci1113: 178–191; Cappello F, Conway de Macario E, Di Felice V, Zummo G, Macario AJL (2009) *Chlamydia trachomatis* infection and anti-Hsp60 immunity: The two sides of the coin. PLoS Pathogens, 5(8): e1000552; doi: 10.1371/journal.ppat.1000552, 2009; and Macario AJL, Cappello F, Zummo G, Conway de Macario E (2010) Chaperonopathies of senescence and the scrambling of the interactions between the chaperoning and the immune systems. Ann New York Acad Sci 1197: 85–93. Specific references pertaining to each cell and effect can be found in these articles and in Further Reading

against it. These would be genuine autoantibodies that will recognize the autologous chaperone, now acting as antigen, and will react with it. This reaction may occur in circulation, in the extracellular space or in the cell surface, depending on where the antigen, i.e., the autoantigen, the autologous chaperone, is located. Disease, i.e., autoimmune disease, may arise associated with the antigen–antibody complexes formed and with lesions predominant in those tissues in which the complexes accumulate.

Another form of interaction between a chaperone and the immune system involves innate immunity components (this chapter). A chaperone binds to the surface of a cell (e.g., neutrophil, macrophage, dendritic cell, monocyte, microglia), via a cell-surface receptor, and thereby triggers production of cytokines and chemokines and starts inflammation. This process may or may not (the point is still under investigation) involve internalization of the chaperone by the target cell, and leads to local lesions in the cells and tissues at which inflammation occurs with subsequent disease.

The third form in which chaperones interact with the immune system is by participating in antigen cross presentation. A chaperone can bind a tumor antigen, forming a chaperone-tumor antigen complex, which is internalized by antigen-presenting cells. The complex can then go either the MHC class I or the MHC class II pathway with activation of T cells. The mechanism of chaperone-tumor antigen complex uptake by antigen-presenting cells can also occur in at least two distinct ways, receptor-mediated or receptor independent. Data indicate that the mechanisms involved are somewhat different from those known to operate in the uptake of antigen not bound to a chaperone, and that chaperone-mediated uptake requires scavenger receptors and proteasome processing leading to complex formation with MHC class I molecules. In summary, chaperone-antigen complex formation, complex uptake and processing by target cells, signaling and ultimate induction of innate and/or adaptive immunity are not yet fully understood for all situations. However, research in this field is very active and should soon provide answers to at least some of questions still open.

It is likely that one or more of the mechanisms described above occur often, thus the lesions and the disease observed are the compound products of innate and adaptive immunities turned against the host driven by a chaperone, a real chaperonopathy by mistake if the culprit is normal and favors disease rather than the contrary. However, one may envision that a chaperone that is recognized as immunogen by its own organism has some abnormality, e.g., a pathological post-translation modification that makes it look foreign to the immune system. In addition, it is not yet clear why a chaperone can elicit a pathological inflammation via interactions with cells of the innate immune system. Do these interactions occur also in physiological conditions, in normal organisms? If so, why at a certain moment in life, or under certain circumstance do they become pathogenic? It could be that during ageing for instance, one or more chaperones suffer aberrant post-translation modifications and, because of this, the interactions with the innate immune system cells are scrambled, which would lead to general inflammation as seen in ageing individuals.

Another factor that might play a determinant role on how the immune system responds to chaperones is the form in which the latter reaches the former, soluble, free, or attached to or packed in exosomes or other extracellular vesicles (see Sect. 2.5 and Chap. 9).

Reports of experimental effects of Hsp60, or any other chaperone, on immune system components (e.g., T- and B-cells, macrophages, monocytes, dendritic cells, neutrophils, microglia, etc.) must be examined with caution because it is often unclear whether the effects observed are due exclusively to the Hsp molecule or to contaminants such as LPS, lipoproteins, and flagellin, or to combinations of them. In our literature searches pertinent to Table 7.3, we focused on reports in which either highly purified human Hsp60 or human Hsp60 peptides that were synthesized *de novo* had been used. Thus, the probability that contaminants were contributing to the effects observed and attributed to Hsp60 was minimized.

Briefly, chaperones have two different sets of functions, depending on whether they are within or outside cells. Outside cells Hsp-chaperones may act as signal molecules producing a "hormonal-like effect" with regard to the immune system. This kind of effect has been suggested for Hsp70, Hsp60, and Hsp10. However, a word of caution is necessary. The non-canonical functions of chaperones are still incompletely understood. Particularly, those functions seemingly due to a "hormonal-like effect," in which extracellular chaperones deliver signals to target cells, are still under scrutiny and are debated. We will have to wait for more data to learn about the reality and extent of these effects and about their intimate mechanism at the molecular level.

7.3.2 Chaperonopathies by Mistake: Chaperones as Autoantigens

Chaperones can be at the center of autoimmune conditions because they are recognized as foreign by the immune system, thus becoming autoantigens. This situation has been described for a number of pathological conditions affecting a variety of tissues and organs. Table 7.4 displays examples Hsp60 locations at which the chaperonin can serve as autoantigen to cause autoimmune lesions and disease together with anti-Hsp60 autoantibodies.

Table 7.4 Hsp60 autoantigen: Locations and pathology

Tissue/organ	Cell/Structure	Pathology
Vessels	Endothelial	Vasculitis, atherosclerosis
Heart	Cardiomyocyte	Myocarditis, infarct, heart failure
Joints	Synoviocyte	Rheumatoid arthritis
Pancreas	Beta	Diabetes
Thyroid	Thyreocyte	Hashimoto's thyroiditis
Liver	Hepatocyte, biliary ducts	Chronic active hepatitis, primary biliary cirrhosis
Adrenal glands	Glomerular-zone	Addison's disease
Nervous System	Synapsis	Myasthenia gravis
Kidney	Endothelial (glomerulus)	Glomerulonephritis
Skin	Keratinocyte, fibroblast, endothelial	Scleroderma, pemphigoid, psoriasis, dermatomyositis
Salivary glands	Not reported	Primary Sjögren's syndrome

Source Macario AJL, Conway de Macario E (2005) Sick chaperones, cellular stress and disease. New Eng J Med 353: 1489–1501; Macario AJL, Conway de Macario E (2007) Chaperonopathies by defect, excess, or mistake. Ann New York Acad Sci 1113: 178–191; Macario AJL, Cappello F, Zummo G, Conway de Macario E (2010) Chaperonopathies of senescence and the scrambling of the interactions between the chaperoning and the immune systems. Ann New York Acad Sci 1197: 85–93; and Cappello F, Conway de Macario E, Di Felice V, Zummo G, Macario AJL (2009) *Chlamydia trachomatis* infection and anti-Hsp60 immunity: The two sides of the coin. PLoS Pathogens, 5(8): e1000552; doi: 10.1371/journal.ppat.1000552, 2009. In these articles and in Further Reading, specific references pertinent to each pathology listed can be found

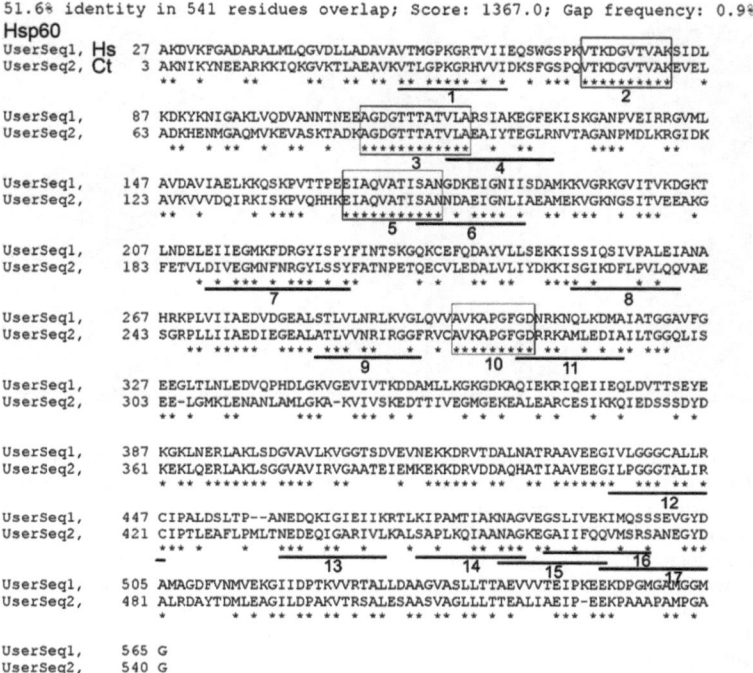

Fig. 7.1 Similarity between human Hsp60 (HsHsp60; Hs stands for *Homo sapiens*) **and** *Chlamydia trachomatis* **Hsp60 (CtHsp60) with shared sequences that could serve as crossreactive epitopes for anti-Hsp60 antibodies elicited by either one of the two proteins.** Pairwise alignment of human (HsHsp60) with *Chlamydia trachomatis* (CtHsp60). Segments with 100 % identity (framed) are 2, 3, 5, and 10. Segments underlined share identities between 33 and 75 %. *Source* Campanella C, Marino Gammazza A, Mularoni L, Cappello F, Zummo G, Di Felice V (2009) A comparative analysis of the products of GROEL-1 gene from *Chlamydia trachomatis* serovar D and the HSP60 var1 transcript from *Homo sapiens* suggests a possible autoimmune response. Intl J Immunogenetics 36: 73–78; and Cappello F, Conway de Macario E, Di Felice V, Zummo G, Macario AJL (2009) *Chlamydia trachomatis* infection and anti-Hsp60 immunity: The two sides of the coin. PLoS Pathogens, 5(8): e1000552; doi: 10.1371/journal. ppat.1000552. For amino acid names and abbreviations see Amino Acid Abbreviations (IUPAC) at http://www.ncbi.nlm.nih.gov/guide/

7.3.3 Possible Reasons why Chaperones Become Autoantigens and, thus, Pathogenic Factors

7.3.3.1 Bacterial Chaperones Elicit Antibodies that Crossreact with the Human Counterparts

We have seen in the preceding section that chaperones, as illustrated by Hsp60, may become autoantigens and cause autoimmune pathology. One possible

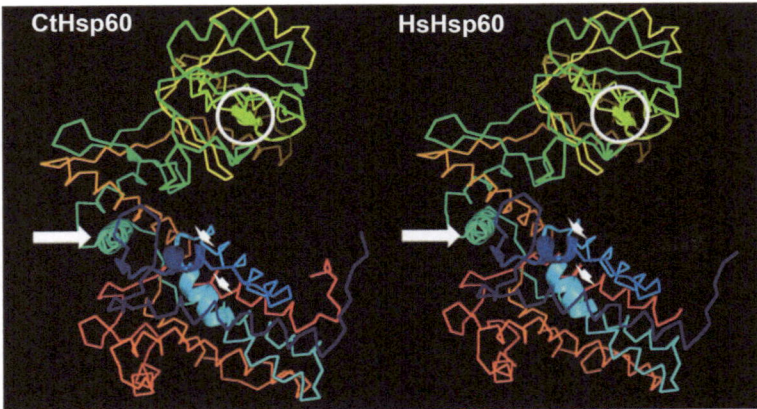

Fig. 7.2 **Structural similarities between human** *(Homo sapiens)* **Hsp60 (variously designated HsHsp60, Hs-Hsp60, or hHSP60) and** *Chlamydia trachomatis* **(CtHsp60, Ct-Hsp60, or ctHSP60) with shared sequences that could function as crossreactive epitopes for anti-Hsp60 antibodies elicited by either one of the two proteins**. See also Table 7.5. The four epitopes with 100 % homologies (framed in Fig. 7.1) are in the blue, light blue, and cyan helixes indicated by the white arrows in the intermediate and equatorial domains, and by the circled light-green arrow in the apical domain. The images were created with PyMol (http://pymol.sourceforge.net). doi:10.1371/journal.ppat.1000552.g001. *Source* Campanella C, Marino Gammazza A, Mularoni L, Cappello F, Zummo G, Di Felice V (2009) A comparative analysis of the products of GROEL-1 gene from *Chlamydia trachomatis* serovar D and the HSP60 var1 transcript from *Homo sapiens* suggests a possible autoimmune response. Intl J Immunogenetics 36:73–78. doi: 10.1111/j.1744-313X.2008.00819.x.; and Cappello F, Conway de Macario E, Di Felice V, Zummo G, Macario AJL (2009) *Chlamydia trachomatis* infection and anti-Hsp60 immunity: The two sides of the coin. PLoS Pathogens 5(8):e1000552. doi:10.1371/journal.ppat.1000552

explanation is that the human chaperone shares potentially immunogenic and antigenic epitopes with the orthologs from pathogens, e.g., bacteria. For example, the Hsp60 from a bacterium harbored by an individual invades the blood and tissues and causes the appearance of antibodies against itself that crossreact with the human counterpart. This is likely the situation for many chronic infections, which may not be symptomatic and, therefore, go untreated for long periods. This silent invasion by a pathogen and its chaperones may lead to an immune reaction against the latter, which ultimately will crossreact with any molecule in the human host that shares epitopes with the bacterial chaperones, including the human chaperones. An example of structural similarities between human and chlamydial Hsp60 that could be the basis of crossreactive antigenic epitopes is shown in Figs. 7.1 and 7.2.

Table 7.5 Primary sequence segments similar in human and bacterial Hsp60 and human acetylcoline receptor alpha 1 (AchRα1)

Sequence segment (aa)	Hsp60	aa number	Similarity %	Out[a]
LIVEKIMQSSSEV	*Homo sapiens*	489–501	53.8	No
FEKISKGANPVEIRRG	(HsHsp60)	128–143	38.9	Yes
VIAELKKQSKPVTTPEE		155–171	35.3	Yes
AIATGGAGEE		317–326	33.3	Yes
AGLLLTTE	*Chlamydia*	512–519	62.5	Yes
GGVAVIRV	*trachomatis*	373–380	50.0	Yes
NIKYNEEARKKIQK	(CtHsp60)	5–18	28.6	Yes
IKYNEEARKKIQKGVKTL		6–23	27.6	Yes
SEQEKLSNY	*Chlamydia*	2–10	55.6	Yes
SEQNQHLIIFCEDID	*pneumoniae*	235–249	53.3	No
DGDAVIAKLSSL	(CpHsp60)	448–459	41.7	Yes
IVKGSYGPKQSLS		26–38	38.5	Yes
YNADKKLFSGIDKLFQIVK		10–28	36.8	Yes
LGVDFAKAMVNKIHKEHS		63–80	33.3	Yes
YENLGVDFAKAMVNKIHKEHSDGAT		60–84	28.0	Yes
PKQSLSPTSFFKERGFYAISQT		33–54	27.3	Yes
LQEALQQQSWPIKDA		121–135	26.7	Yes
LHAILQESYAALEKGISTHKLIASLKLQGEKLQ		90–122	18.2	Yes

[a] Sequence segment (epitope) location. Out, outside of the molecule, i.e., exposed to the surface
Source Marino Gammazza A, Bucchieri F, Grimaldi LM, Benigno A, Conway de Macario E, Macario AJL, Zummo G, Cappello F (2012) The molecular anatomy of human Hsp60 and its similarity with that of bacterial orthologs and acetylcholine receptor reveal a potential pathogenetic role of anti-chaperonin immunity in myasthenia gravis. Cell Mol Neurobiol 32:943–947. DOI: 10.1007/s10571-011-9789-8. For amino acid names and abbreviations see Amino Acid Abbreviations (IUPAC) at http://www.ncbi. nlm.nih.gov/guide/)

7.3.3.2 Bacterial Chaperones Elicit Antibodies that Crossreact also with Human Molecules Other than Chaperones and that may Cause Autoimmunity

The sharing of epitopes between human and bacterial Hsp60 and other human molecules (e.g., acetylcholine receptors, thyroglobulin, and thyroperoxidase) makes it possible the appearance of crossreactive antibodies reacting with human Hsp60 and other human molecules. These crossreactive antibodies may very well be pathogenic factors in a number of autoimmune conditions such as myasthenia gravis (Table 7.5 and Fig. 7.3), and Hashimoto's thyroiditis.

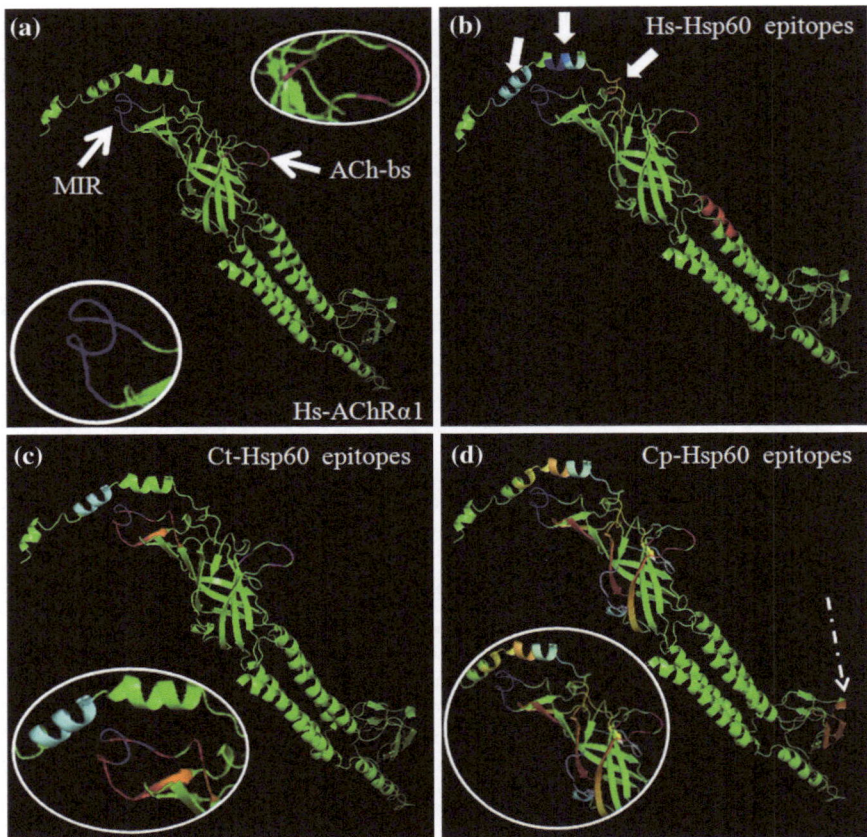

Fig. 7.3 Three-dimensional model of AChRa1 monomer showing shared amino-acid sequences (potential immunogenic-antigenic epitopes) with the human Hsp60 (Hs-Hsp60) and its chlamydial counterparts (see also Fig. 7.1, and Table 7.5). **a** Shown are the MIR (Main Immunogenic Region; blue loop, bottom left corner insert) and the ACh-binding site (ACh-bs; hot pink discontinuous loop, top right corner insert). **b** Homologous epitopes revealed by the alignment of AChRa1 and Hs-Hsp60 sequences. One epitope maps to the transmembrane region of AChRa1 (red helix, aa 489–501), while the other three epitopes (shown by white thick arrows) map to around the MIR. **c** Homologous epitopes revealed by the alignment of AChRa1 and *Chlamydia trachomatis* Hsp60 (Ct-Hsp60) sequences. Four epitopes (cyan helix and red loops- see bottom left corner insert) map around and on to the MIR. **d** Homologous epitopes revealed by the alignment of AChRa1 and *Chlamydia pneumoniae* Hsp60 (Cp-Hsp60) sequences. Of the 10 shared sequences, only one (shown by a white discontinuous arrow, aa 235–249) maps to the intracellular region of AChRa1, whereas all the rest (various colors; see insert down to the left) map around and on the ACh-binding site. *Source* Marino Gammazza A, Bucchieri F, Grimaldi LM, Benigno A, Conway de Macario E, Macario AJL, Zummo G, Cappello F (2012) The molecular anatomy of human Hsp60 and its similarity with that of bacterial orthologs and acetylcholine receptor reveal a potential pathogenetic role of anti-chaperonin immunity in myasthenia gravis. Cell Mol Neurobiol 32:943–947. doi:10.1007/s10571-011-9789-8

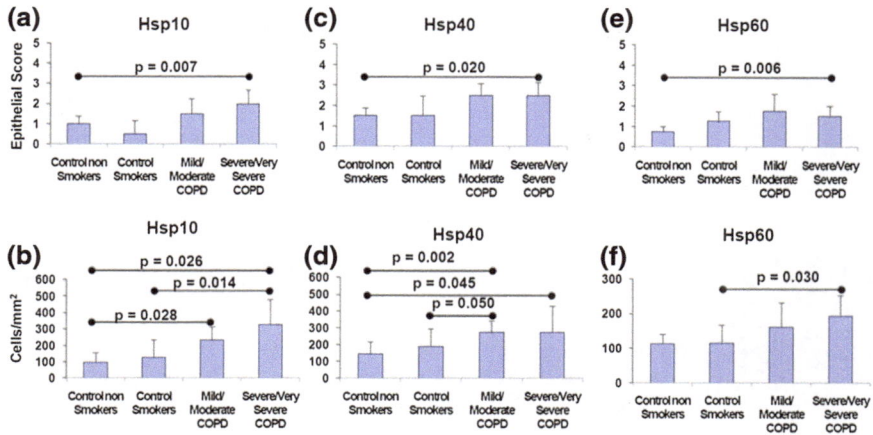

Fig. 7.4 Measurement of Hsps in the bronchial epithelium and lamina propria: Comparison between groups of patients. Hsp10, Hsp40, and Hsp60 immuno positive cells in the epithelium (*top panels* **a, c, e**) and in the lamina propria (*bottom panels* **b, d, f**) of the four groups studied: Control non Smokers, Control Smokers, Mild/Moderate COPD, and Severe/Very Severe COPD. Results are expressed as the median and interquartile range (*IQR*) of scored (*0–3, vertical axis*) immunopositivity in the epithelium, or as number of immuno positive cells per square millimeter (*vertical axis*) in the lamina propria; "p" values are shown on top of the lines spanning the two groups being compared. *Source* Cappello F, Caramori G, Campanella C, Vicari C, Gnemmi I, Zanini A, Spanevello A, Capelli A, La Rocca G, Anzalone R, Bucchieri F, D'Anna SE, Ricciardolo FLM, Brun P, Balbi B, Carone M, Zummo G, Conway de Macario E, Macario AJL, Di Stefano A (2011). Convergent sets of data from in vivo and in vitro methods point to an active role of Hsp60 in chronic obstructive pulmonary disease pathogenesis. PLoS ONE 6(11): e28200, 2011

7.4 Chaperones and Inflammation

Chaperones can act as inducers of inflammation by stimulating cells of the innate immune system to produce inflammatory cytokines and, thus, initiate and/or maintain chronic inflammatory diseases. Examples of this situation follow.

7.4.1 Chronic Obstructive Pulmonary Disease

There are increased levels of Hsp10, Hsp40, and Hsp60 in the mucosa of airways in COPD, and the magnitude of the increase correlates positively with the severity of the pathology (Figs. 7.4 and 7.5). This pattern suggests that the quantities of the chaperones and the lesions are closely related and gives support to the notion that the chaperones are involved in pathogenesis, considering what we know now on the pro-inflammatory action that the chaperones can have.

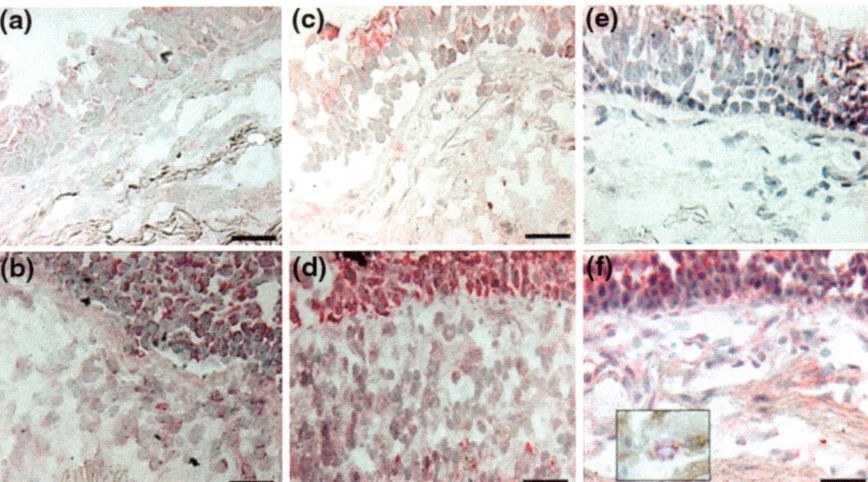

Fig. 7.5 Hsps in the bronchial epithelium and lamina propria: Representative images. Photomicrographs showing frozen sections of bronchial mucosa from a control nonsmoker (**a, c, e**) and from a patient with severe stable COPD (**b, d, f**) immunostained to identify Hsp10 (**a, b**), Hsp40 (**c, d**), and Hsp60 (**e, f**). Nuclei were counterstained with haematoxylin (*blue*). Cells positive for Hsps are in *red*. Inset in **f** shows a cell double stained for neutrophil elastase (*red*) and Hsp60 (*brown*). Bar = 50 microns. *Source* Cappello F, Caramori G, Campanella C, Vicari C, Gnemmi I, Zanini A, Spanevello A, Capelli A, La Rocca G, Anzalone R, Bucchieri F, D'Anna SE, Ricciardolo FLM, Brun P, Balbi B, Carone M, Zummo G, Conway de Macario E, Macario AJL, Di Stefano A (2011). Convergent sets of data from in vivo and in vitro methods point to an active role of Hsp60 in chronic obstructive pulmonary disease pathogenesis. PLoS ONE 6(11): e28200, 2011

Further support to the notion that Hsp60 interacts with inflammatory cells and, thereby, have pathogenic potential in COPD, derives from the observed strict correlation and co-localization of Hsp60 with neutrophils and, very suggestive, with activated neutrophils (Fig. 7.6).

7.4.2 Inflammatory Bowel Disease: Ulcerative Colitis

Chronic inflammation of the intestinal mucosa is prominent among the lesions that characterize UC. The role of inflammatory cells in pathogenesis is indicated by their increase in diseased intestinal mucosa and by their decrease in response to efficacious treatments leading to patient's improvement. It is noteworthy that the more efficacious the treatment the greater the decrease in inflammatory cells, Fig. 7.7.

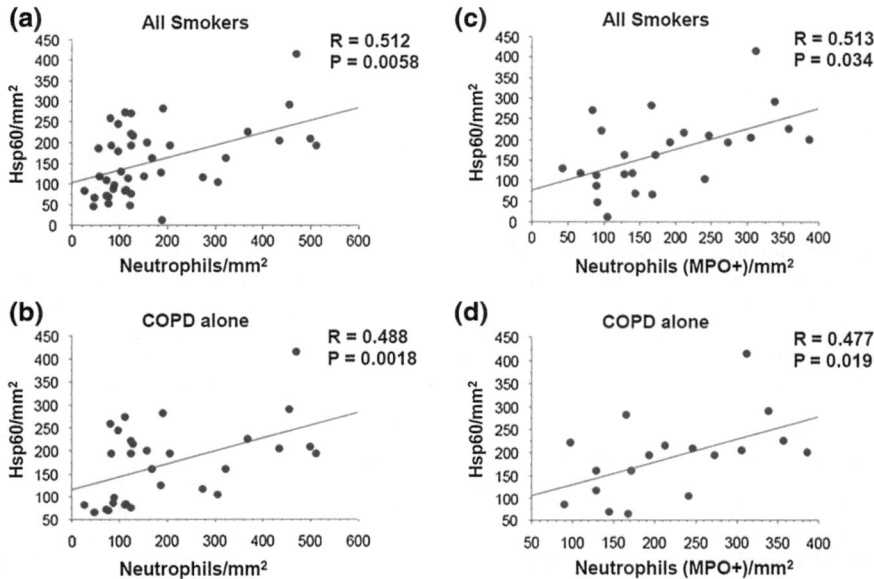

Fig. 7.6 Correlations between neutrophils and Hsp60. Regression analysis between number of Hsp60 positive cells (*vertical axis*) and number of neutrophils (*horizontal axis*), panels **a** and **b**, and between number of Hsp60 positive cells (*vertical axis*) and number of MPO positive neutrophils (*horizontal axis*), panels **c** and **d**, in the lamina propria of all smokers (panels **a** and **c**) and of patients with COPD alone, considered as a single group (panels **b** and **d**). *Source* Cappello F, Caramori G, Campanella C, Vicari C, Gnemmi I, Zanini A, Spanevello A, Capelli A, La Rocca G, Anzalone R, Bucchieri F, D'Anna SE, Ricciardolo FLM, Brun P, Balbi B, Carone M, Zummo G, Conway de Macario E, Macario AJL, Di Stefano A (2011). Convergent sets of data from in vivo and in vitro methods point to an active role of Hsp60 in chronic obstructive pulmonary disease pathogenesis. PloS ONE 6(11): e28200, 2011

Likewise, the notion that Hsp60 plays a pathogenic role together with inflammatory cells, perhaps *depending* on its interaction with inflammatory cells, in UC is supported by the fact that the levels of the chaperonin in the affected mucosa follow the same trend as inflammatory cells: increased before treatment and decreased after it, the decrease in Hsp60 levels being greater in patients most improved with the least inflammatory cells in their intestinal mucosae (Fig. 7.8). This is further illustrated by the observed positive, significant correlation between Hsp60 and inflammatory cells, Fig. 7.9. Finally, a direct association of Hsp60 and inflammatory cells is demonstrated by co-localization tests, which showed, for instance, co-localization of Hsp60 and CD68-bearing inflammatory cells, Fig. 7.10.

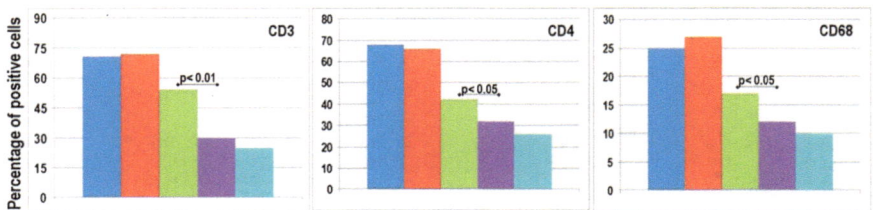

Fig. 7.7 Immunohistochemical results for inflammatory markers. The bars show the levels of *CD3*, *CD4*, and *CD68* in biopsies from patients before and after treatment with either 5-ASA alone or with the combination 5-ASA + PB. Both treatments produced a significant reduction ($P < 0.01$) in the percentage of inflammatory cells. The reduction of inflammatory cells after treatment with 5-ASA + PB was significantly greater ($P < 0.01$, $P < 0.05$, and $P < 0.05$ for, respectively, *CD3*, *CD4*, and *CD68*) than after treatment with 5-ASA alone. *Bars*: *dark blue*, 5-ASA at the time of diagnosis; *red*, 5-ASA + PB at the time of diagnosis; *green*, 5-ASA after 6 months; *violet*, 5-ASA + PB after 6 months; *light blue*, normal mucosa (controls). 5-ASA indicates 5-aminosalicylic acid (mesalazine); PB, probiotics. *Source* Tomasello G, Rodolico V, Zerilli M, Martorana A, Bucchieri F, Pitruzzella A, Gammazza AM, David S, Rappa F, Zummo G, Damiani P, Accomando S, Rizzo M, Conway de Macario E, Macario AJL, Cappello F (2011) Changes in immunohistochemical levels and subcellular localization after therapy and correlation and colocalization with CD68 suggest a pathogenetic role of Hsp60 in ulcerative colitis. Appl Immunohistochem Mol Morphol 19: 552–561

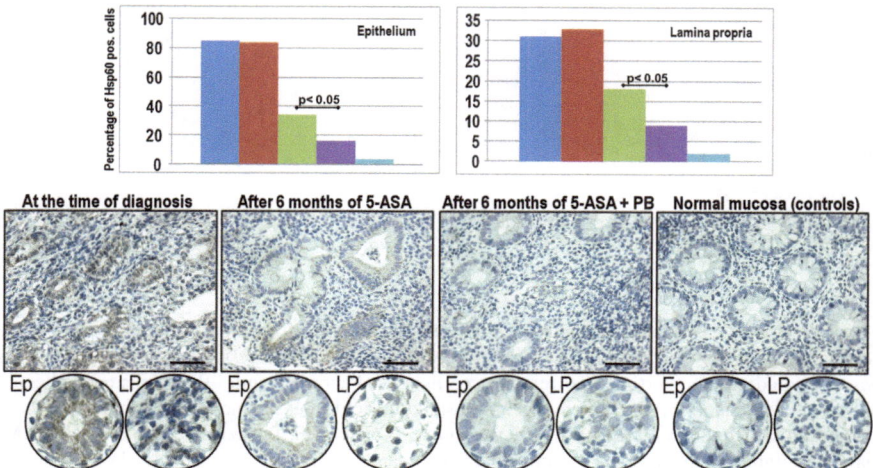

Fig. 7.8 Immunohistochemical results and representative images of Hsp60-positive cells in colon mucosa. The top 2 panels display bars showing the levels of Hsp60-positive cells in epithelium and lamina propria. Both treatments determined a significant reduction of Hsp60 levels ($P < 0.01$). Moreover, Hsp60 reduction in 5-ASA + PB treated patients was significantly greater ($P < 0.05$) than in patients receiving 5-ASA alone. The middle panels show representative images of the immunohistochemical results for Hsp60. Positive cells are brown. Nuclei were counterstained with hematoxylin and appear blue. Bar: approximately 100 microns. The bottom circular panels show at a magnification higher than that in the middle panels the immunohistochemical images for Hsp60-positive cells in epithelium (*Ep*) and lamina propria (*LP*), demonstrating that the immuno positivity was localized in the cytoplasm. *Top panel bars*: *dark blue*, 5-ASA at the time of diagnosis; *red*, 5-ASA + PB at the time of diagnosis; *green*, 5-ASA after 6 months; *violet*, 5-ASA + PB after 6 months; *light blue*, normal mucosa (controls). 5-ASA indicates 5-aminosalicylic acid (mesalazine); PB, probiotics. *Source* Tomasello G, Rodolico V, Zerilli M, Martorana A, Bucchieri F, Pitruzzella A, Gammazza AM, David S, Rappa F, Zummo G, Damiani P, Accomando S, Rizzo M, Conway de Macario E, Macario AJL, Cappello F (2011) Changes in immunohistochemical levels and subcellular localization after therapy and correlation and colocalization with CD68 suggest a pathogenetic role of Hsp60 in ulcerative colitis. Appl Immunohistochem Mol Morphol 19: 552–561

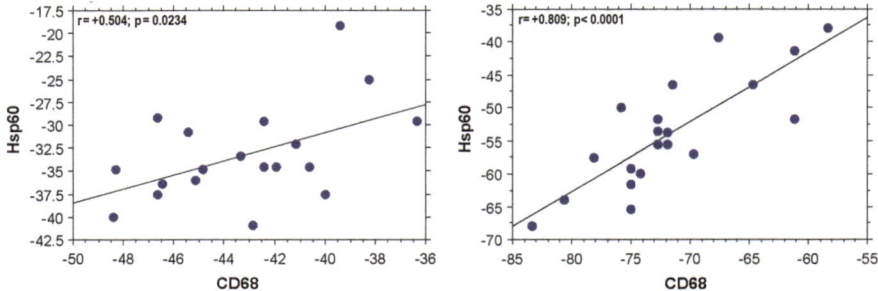

Fig. 7.9 Hsp60-CD68 positive correlation. The regression plots show a significant positive correlation between Hsp60 and CD68 levels in both 5-ASA (*left panel*) and 5-ASA + PB (*right panel*) treated groups. Pertinent correlation coefficients (Pearson r) and "P" values are indicated on *top* of each panel. 5-ASA indicates 5-aminosalicylic acid (mesalazine); PB, probiotics. *Source* Tomasello G, Rodolico V, Zerilli M, Martorana A, Bucchieri F, Pitruzzella A, Gammazza AM, David S, Rappa F, Zummo G, Damiani P, Accomando S, Rizzo M, Conway de Macario E, Macario AJL, Cappello F (2011) Changes in immunohistochemical levels and subcellular localization after therapy and correlation and colocalization with CD68 suggest a pathogenetic role of Hsp60 in ulcerative colitis. Appl Immunohistochem Mol Morphol 19: 552–561

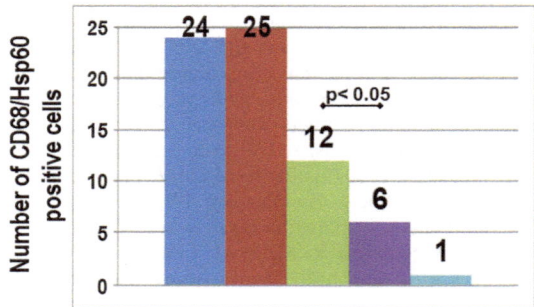

Fig. 7.10 Double immunofluorescence results for Hsp60 and CD68. The bars show the mean number of cells showing Hsp60-CD68 colocalization per high power field (*400*). This number was reduced significantly after both treatments ($P < 0.01$) and the reduction after 5-ASA + PB treatment was significantly greater ($P < 0.05$) than after treatment with 5-ASA alone. Hsp60 and CD68 often colocalized in all specimens but the greatest numbers of cells showing Hsp60-CD68 colocalization were recorded in the samples taken before treatment. *Colors: dark blue*, 5-ASA at the time of diagnosis; *red*, 5-ASA + PB at the time of diagnosis; *green*, 5-ASA after 6 months; *violet*, 5-ASA + PB after 6 months; *light blue*, normal mucosa (controls). 5-ASA indicates 5-aminosalicylic acid (mesalazine); PB, probiotics. *Source* Tomasello G, Rodolico V, Zerilli M, Martorana A, Bucchieri F, Pitruzzella A, Gammazza AM, David S, Rappa F, Zummo G, Damiani P, Accomando S, Rizzo M, Conway de Macario E, Macario AJL, Cappello F (2011) Changes in immunohistochemical levels and subcellular localization after therapy and correlation and colocalization with CD68 suggest a pathogenetic role of Hsp60 in ulcerative colitis. Appl Immunohistochem Mol Morphol 19: 552–561

Further Reading

Section 7.1

Chaperone Misplacement

Becker D, Krayl M, Strub A, Li Y, Mayer MP, Voos W (2008) Impaired interdomain communication in mitochondrial Hsp70 results in the loss of inward-directed translocation force. J Biol Chem 284:2934–2946

Ribeil JA, Zermati Y, Vandekerckhove J, Cathelin S, Kersual J, Dussiot M, Coulon S, Moura IC, Zeuner A, Kirkegaard-Sørensen T, Varet B, Solary E, Garrido C, Hermine O (2007) Hsp70 regulates erythropoiesis by preventing caspase-3-mediated cleavage of GATA-1. Nature 445:102–105

Section 7.2

QUANTITATIVE ABNORMALITIES OF CHAPERONES AND CHAPERONOPATHIES BY MISTAKE OR COLLABORATIONISM: CANCER HYOU-1
HYOU in Cancer Invasion

Stojadinovic A, Hooke JA, Shriver CD, Nissan A, Kovatich AJ, Kao T-C, Ponniah S, Peoples GE, Moroni M (2007) HYOU1/Orp150 expression in breast cancer. Med Sci Monit 13:BR231–239 17968289 (P, S, E, B)

HYOU-1 in Tumor Angiogenesis

It was demonstrated in vitro with cell cultures that ORP150 plays a determinant role in tumor-mediated angiogenesis via VEGF processing

Ozawa K, Tsukamoto Y, Hori O, Kitao Y, Yanagi H, Stern DM, Ogawa S (2001) Regulation of tumor angiogenesis by oxygen-regulated protein 150, an inducible endoplasmic reticulum chaperone. Cancer Res 61:4206–4213

ORP150 in the Pathogenesis of Cancer and Other, ER Stress-Related, Disorders: Diabetes, and Neurodegenerative and Cardiovascular Diseases

Kusaczuk M, Cechowska-Pasko M (2013) Molecular chaperone ORP150 in ER stress-related diseases. Curr Pharm Des. 2013 Jan 29. [Epub ahead of print]

Hsp60 and Hsp10: Hsp60 and Apopotosis

Chandra D, Choy G, Tang DG (2007) Cytosolic accumulation of HSP60 during apoptosis with or without apparent mitochondrial release—evidence that its pro-apoptotic or pro-survival functions involve differential interactions with caspase-3. J Biol Chem 282:31289–31301

Hsp60 and Metastasis

Tsai YP, Yang MH, Huang CH, Chang SY, Chen PM, Liu CJ, Teng SC, Wu KJ (2009) Interaction between HSP60 and beta-catenin promotes metastasis. Carcinogenesis 30:1049–1057

Hsp60 Increased in Breast Cancer

Desmetz C, Bibeau F, Boissière F, Bellet V, Rouanet P, Maudelonde T, Mangé A, Solassol J (2008) Proteomics-based identification of HSP60 as a tumor-associated antigen in early stage breast cancer and ductal carcinoma in situ. J Proteome Res 7:3830–3837

Hsp60 in Uterine Cervix Carcinoma

Cappello F, Bellafiore M, Palma A, Marciano V, Martorana G, Belfiore P, Martorana A, Farina F, Zummo G, Bucchieri F (2002–2003). Expression of 60-kD heat shock protein increases during carcinogenesis in the uterine exocervix. Pathobiology 70:83–88

Cappello F, Bellafiore M, David S, Anzalone R, Zummo G (2003) Ten kilodalton heat shock protein (HSP10) is overexpressed during carinogenesis of large bowel and uterine exocervix. Cancer Lett 196:35–41

Hwang YJ, Lee SP, Kim SY, Choi YH, Kim MJ, Lee CH, Lee JY, Kim DY (2009) Expression of heat shock protein 60 kDa is upregulated in cervical cancer. Yonsei Med J 50:399–406

Hsp60 and Others in Prostate Cancer

Cappello F, Rappa F, David S, Anzalone R, Zummo G (2003) Immunohistochemical evaluation of PCNA, p53, HSP60, HSP10 and MUC-2 presence and expression in prostate carcinogenesis. Anticancer Res 23(2B):1325–1331

Castilla C, Congregado B, Conde JM, Medina R, Torrubia FJ, Japón MA, Sáez C (2010) Immunohistochemical expression of Hsp60 correlates with tumor progression and hormone resistance in prostate cancer. Urology 76(4):1017.e1–6

Glaessgen A, Jonmarker S, Lindberg A, Nilsson B, Lewensohn R, Ekman P, Valdman A, Egevad L (2008) Heat shock proteins 27, 60 and 70 as prognostic markers of prostate cancer. APMIS 116:888–895

Hsp70
Hsp70-2 in Urothelial Cancer

Garg M, Kanojia D, Seth A, Kumar R, Gupta A, Surolia A, Suri A (2010a) Heat-shock protein 70–2 (HSP70-2) expression in bladder Urothelial carcinoma is associated with tumour progression and promotes migration and invasion. Eur J Cancer 46:207–215

Hsp72 in Chronic Lymphocytic and Myelomonocytic Leukemias

Madden LA, Hayman YA, Underwood C, Vince RV, Greenman J, Allsup D, Ali S (2012) Increased inducible heat shock protein 72 expression associated with PBMC isolated from patients with haematological tumours. Scand J Clin Lab Invest 72:380–386

Hsp72 and Hsp73 Together with Hsp90 Favor Multiple Myeloma

Chatterjee M, Andrulis M, Stühmer T, Müller E, Hofmann C, Steinbrunn T, Heimberger T, Schraud H, Kressmann S, Einsele H, Bargou RC (2012) The PI3 K/Akt signalling pathway regulates the expression of Hsp70, which critically contributes to Hsp90-chaperone function and tumor cell survival in multiple myeloma. Haematologica. doi:10.3324/haematol.2012. 066175

Post-Translationaly Modified Hsp70 Favors Cancer

Cho HS, Shimazu T, Toyokawa G, Daigo Y, Maehara Y, Hayami S, Ito A, Masuda K, Ikawa N, Field HI, Tsuchiya E, Ohnuma S, Ponder BA, Yoshida M, Nakamura Y, Hamamoto R (2012) Enhanced HSP70 lysine methylation promotes proliferation of cancer cells through activation of Aurora kinase B. Nat Commun 18(3):1072. doi:10.1038/ncomms2074

Hsp70 Favors Tumors Instead of Fighting Against Them

Boroughs LK, Antonyak MA, Johnson JL, Cerione RA (2011) A unique role for heat shock protein 70 and its binding partner tissue transglutaminase in cancer cell migration. J Biol Chem 286:37094–37107

Garg M, Kanojia D, Saini S, Suri S, Gupta A, Surolia A, Suri A (2010b) Germ cell-specific heat shock protein 70-2 is expressed in cervical carcinoma and is involved in the growth, migration, and invasion of cervical cells. Cancer 116:3785–3796

Li G, Xu Y, Guan D, Liu Z, Liu DX (2011a) HSP70 promotes survival of C6 and U87 glioma cells by inhibition of ATF5 degradation. J Biol Chem 286:20251–20259

Macario AJL, Cappello F, Conway de Macario E (2012) Chaperonopathies: diseases in which mortalin and other Hsp-chaperones play a role in etiology and pathogenesis. In: Kaul SC, Wadhwa R (eds) Mortalin biology: life, stress and death. Springer, Berlin, pp 209–221

Nylandsted J, Brand K, Jäättelä M (2000) Heat shock protein 70 is required for the survival of cancer cells. Ann New York Acad Sci 926:122–125

Rérole AL, Jego G, Garrido C (2011) Hsp70: Anti-apoptotic and tumorigenic protein. Methods Mol Biol 787:205–230

Rohde M, Daugard M, Jensen MH, Helin K, Nylandsted J, Jaattela M (2005) Members of the heat-shock protein 70 family promote cancer cell growth by distinct mechanisms. Genes Dev 19:570–582

Teng Y, Ngoka L, Mei Y, Lesoon L, Cowell JK (2012) HSP90 and HSP70 are essential for stabilization and activation of the WASF3 metastasis promoting protein. J Biol Chem 287:10051–10059

BAG3; BAG5

Boiani M, Daniel C, Liu X, Hogarty MD, Marnett LJ (2013) The stress protein BAG3 stabilizes Mcl-1 and promotes survival of cancer cells and resistance to ABT-737. J Biol Chem 288:6980–6990. doi:10.1074/jbc.M112.414177

Bruchmann A, Roller C, Walther TV, Schäfer G, Lehmusvaara S, Visakorpi T, Klocker H, Cato AC, Maddalo D (2013) Bcl-2 associated athanogene 5 (Bag5) is overexpressed in prostate cancer and inhibits ER-stress induced apoptosis. BMC Cancer 13(1):96

Festa M, Del Valle L, Khalili K, Franco R, Scognamiglio G, Graziano V, De Laurenzi V, Turco MC, Rosati A (2011) BAG3 protein is overexpressed in human glioblastoma and is a potential target for therapy. Am J Pathol 178:2504–2512

Hsp90
Hsp90 and Breast Cancer

Pick E, Kluger Y, Giltnane JM, Moeder C, Camp RL, Rimm DL, Kluger HM (2007) High HSP90 expression is associated with decreased survival in breast cancer. Cancer Res 67:2932–2937

Hsp90 Increased in Pancreatic Endocrine Tumors (PET)

Mayer P, Harjung A, Breinig M, Fischer L, Ehemann V, Malz M, Scherübl H, Britsch S, Werner J, Kern MA, Bläker H, Schirmacher P, Bergmann F (2012) Expression and therapeutic relevance of heat shock protein 90 in pancreatic endocrine tumors. Endocr Relat Cancer 19:217–232

Hsp70 and Hsp90 are Increased Along with HOP and Decreased CHIP in Colon Cancer

Ruckova E, Muller P, Nenutil R, Vojtesek B (2012) Alterations of the Hsp70/Hsp90 chaperone and the HOP/CHIP co-chaperone system in cancer. Cell Mol Biol Lett 17:446–458

Hsp90 Plus CDC37 Protect the Oncogenic Protein FOP2-FGFR1 in Myeloid Leukemia

Jin Y, Zhen Y, Haugsten EM, Wiedlocha A (2011) The driver of malignancy in KG-1a leukemic cells, FGFR1OP2-FGFR1, encodes an HSP90 addicted oncoprotein. Cell Signal 23:1758–1766

Hsp90 Co-chaperone p23

Simpson NE, Lambert WM, Watkins R, Giashuddin S, Huang SJ, Oxelmark E, Arju R, Hochman T, Goldberg JD, Schneider RJ, Reiz LF, Soares FA, Logan SK, Garabedian MJ (2010) High levels of Hsp90 cochaperone p 23 promote tumor progression and poor prognosis in breast cancer by increasing lymph node metastases and drug resistance. Cancer Res 70:8446–8456

Hsp90 in Myelodysplastic Syndrome-acute Myeloid Leukemia

Flandrin-Gresta P, Solly F, Aanei CM, Cornillon J, Tavernier E, Nadal N, Morteux F, Guyotat D, Wattel E, Campos L (2012) Heat shock protein 90 is overexpressed in high-risk myelodysplastic syndromes and associated with higher expression and activation of focal adhesion kinase. Oncotarget 3:1158–1168

Hsp90 Favors Prostate Cancer Progression and Metastasis

Hance MW, Dole K, Gopal U, Bohonowych JE, Jezierska-Drutel A, Neumann CA, Liu H, Garraway IP, Isaacs JS (2012) Secreted Hsp90 is a Novel Regulator of the Epithelial to Mesenchymal Transition (EMT) in Prostate Cancer. J Biol Chem 287:37732–37744

Pin1 and Non-small Cell Lung Carcinoma (NSCLC)

Tan Xiaogang, Zhou Fang, Wan Junting, Hang Jie, Chen Zhaoli, Li Baozhong, Zhang Cuiyan, Shao Kang, Jiang Peng, Shi Susheng, Feng Xiaoli, Lv Ning, Wang Zhen, Ling Yun, Zhao Xiaohong, Ding Dapeng, Sun Jian, Xiong Meihua, Jie He (2010) Pin1 expression contributes to lung cancer prognosis and carcinogenesis. Cancer Biol Ther 9:111–119

Qian-Lin Zhu, Ting-Feng Wang, Qi-Feng Cao, Min-Hua Zheng, Ai-Guo Lu (2010) Inhibition of cytosolic chaperonin CCTζ-1 expression depletes proliferation of colorectal carcinoma in vitro. J Surg Oncol 102:419–423

MITOCHONDRIAL CHAPERONES AND CANCER
Trap1

Landriscina M, Laudiero G, Maddalena F, Amoroso MR, Piscazzi A, Cozzolino F, Monti M, Garbi C, Fersini A, Pucci P, Esposito F (2010) Mitochondrial chaperone Ttrap1 and the calcium binding protein sorcin interact and protect cells against apoptosis induced by antiblastic agents. Cancer Res 70:6577–6586

HSPA9-B (Mortalin)

Dundas SR, Lawrie LC, Rooney PH, Murray GI (2005) Mortalin is over-expressed by colorectal adenocarcinomas and correlates with poor survival. J Pathol 205:74–81

Wadhwa R, Takano S, Kaur K, Deocaris CC, Pereira-Smith OM, Reddel RR, Kaul SC (2006) Upregulation of mortalin/mthsp70/Grp75 contributes to human carcinogenesis. Int J Cancer 118:2973–2980

ER CHAPERONES AND CANCER

Luo B, Lee AS (2012) The critical roles of endoplasmic reticulum chaperones and unfolded protein response in tumorigenesis and anticancer therapies. Oncogene 2012:1–14. doi:10.1038/onc.2012.130

Grp78 (BiP; HSPA5)
Grp78 is Increased in Ovarian Cancer Cells and Expressed on the Surface of These Cells and May Contribute to Tumor Expansion

Delie F, Petignat P, Cohen M (2012) GRP78 protein expression in ovarian cancer patients and perspectives for a drug-targeting approach. J Oncol 2012:468615. doi:10.1155/2012/468615

Overexpression of Grp78 Indicates Poor Prognosis in Gastric Carcinomas

Zheng HC, Takahashi H, Li XH, Hara T, Masuda S, Guan YF, Takano Y (2008) Overexpression of GRP78 and GRP94 are markers for aggressive behavior and poor prognosis in gastric carcinomas. Hum Pathol 39:1042–1049

EPIGENETIC REGULATION OF CHAPERONE GENES IN CANCER AND OTHER CHAPERONOPATHIES

Krøll J (2007) Molecular Chaperones and the Epigenetics of Longevity and Cancer Resistance. Ann New York Acad Sci 1100:75–83

Krøll J (2010) Correlations of plasma cortisol levels, chaperone expression and mammalian longevity: a review of published data. Biogerontology 11:495–499

Martínez Picabea de Giorgiutti E (2011) On chaperones, epigenesis and disease. Medicina (Buenos Aires) 71:302–303

Mi R, Song L, Wang Y, Ding X, Zeng J, Lehoux S, Aryal RP, Wang J, Crew VK, van Die I, Chapman AB, Cummings RD, Ju T (2012) Epigenetic silencing of the chaperone Cosmc in human leukocytes expressing Tn antigen. J Biol Chem 287:41523–41533

Serrano A, Redondo M, Tellez T, Castro-Vega I, Roldan MJ, Mendez R, Rueda A, Jimenez E (2009) Regulation of Clusterin expression in human cancer via DNA methylation. Tumor Biol 30:286–291

Waha A, Felsberg J, Hartmann W, Hammes J, Knesebeck AV, Endl E, Pietsch T, Waha A (2011) Frequent epigenetic inactivation of the chaperone SGNE1/7B2 in human gliomas. Int J Cancer 131:612–622

Zhang Y, Wang R, Song H, Huang G, Yi J, Zheng Y, Wang J, Chen L (2011) Methylation of multiple genes as a candidate biomarker in non-small cell lung cancer. Cancer Lett 303:21–28

Zhou P, Luo Y, Liu X, Fan L, Lu Y (2012) Down-regulation and CpG island hypermethylation of CRYAA in age-related nuclear cataract. FASEB J 26:4897–4902. doi:10.1096/fj.12-213702

Cosmc Favors Colon Cancer
Huang J, Che MI, Lin NY, Hung JS, Huang YT, Lin WC, Huang HC, Lee PH, Liang JT, Huang
    MC (2013) The molecular chaperone Cosmc enhances malignant behaviors of colon cancer
    cells via activation of Akt and ERK. Mol Carcinog. doi:10.1002/mc.22011

INTERPLAY BETWEEN THE CHAPERONING AND UBIQUITIN-PROTEASOME SYSTEMS, AUTOPHAGY, AND APOPTOSIS

Benbrook DM, Long A (2012) Integration of autophagy, proteasomal degradation, unfolded
    protein response and apoptosis. Exp Oncol 34:286–297

Sections 7.3–7.4

ILLUSTRATIVE EXAMPLE OF A CHAPERONE WITH CANONICAL (PERTAINING TO PROTEIN FOLDING) AND NON-CANONICAL FUNCTIONS

McGreal RS, Brennan LA, Kantorow WL, Wilcox JD, Wei J, Chauss D, Kantorow M (2013)
    Chaperone-independent mitochondrial translocation and protection by αB-crystallin in RPE
    cells. Exp Eye Res 2013 Mar 4. pii: S0014–4835(13):00052–3. doi:10.1016/j.exer.2013.02.016

THE NON-CHAPERONING ROLES OF CHAPERONES: INTERACTION WITH THE IMMUNE SYSTEM. AUTOIMMUNITY, INFLAMMATION
Hsp are Signaling (Hormone-Like) Molecules with Impact on the Immune System or, in Other Words, Hsp can Act as Extracellular Messengers Targeting Immune System Cells. Participation of Hsp in Antigen Presentation
Asea A (2008) Hsp70: a chaperokine. Novartis Found Symp 291:173–179; discussion 179–83,
    221–224
Asea A, Kraeft SK, Kurt-Jones EA, Stevenson MA, Chen LB, Finberg RW, Koo GC, Calderwood
    SK (2000) HSP70 stimulates cytokine production through a CD14-dependant pathway,
    demonstrating its dual role as a chaperone and cytokine. Nat Med 6:435–442
Baranova IN, Vishnyakova TG, Bocharov AV, Leelahavanichkul A, Kurlander R, Chen Z, Souza
    AC, Yuen PS, Star RA, Csako G, Patterson AP, Eggerman TL (2012) Class B scavenger
    receptor types I and II and CD36 mediate bacterial recognition and proinflammatory signaling
    induced by *Escherichia coli*, lipopolysaccharide, and cytosolic chaperonin 60. J Immunol
    188:1371–1380
Basu S, Binder RJ, Suto R, Anderson KM, Srivastava PK (2000) Necrotic but not apoptotic cell
    death releases heat shock proteins, which deliver a partial maturation signal to dendritic cells
    and activate the NF-kappa B pathway. Int Immunol 12:1539–1546
Bethke K, Staib F, Distler M, Schmitt U, Jonuleit H, Enk AH, Galle PR, Heike M (2002)
    Different efficiency of heat shock proteins (HSP) to activate human monocytes and dendritic
    cells: superiority of HSP60. J Immunol 169:6141–6188
Calderwood SK, Mambula SS, Gray PJ Jr (2007) Extracellular heat shock proteins in cell
    signaling and immunity. Ann New York Acad Sci 1113:28–39
Chen T, Cao X (2010) Stress for maintaining memory: HSP70 as mobile messenger for innate
    immunity and adaptive immunity. Eur J Immunol 40:1541–1544
Habich C, Burkart V (2007) Heat shock protein 60: regulatory role on innate immune cells. Cell
    Mol Life Sci 64:742–751

Henderson B (2009) Integrating the cell stress response: a new view of molecular chaperones as immunological and physiological homeostatic regulators. Cell Biochem Funct 28:1–14

Henderson B, Calderwood SK, Coates AR, Cohen I, van Eden W, Lehner T, Pockley AG (2010) Caught with their PAMPs down? The extracellular signalling actions of molecular chaperones are not due to microbial contaminants. Cell Stress Chaperones 15:123–141

Multhoff G (2009) Activation of natural killer cells by heat shock protein 70. Int J Hyperthermia 25:169–175

Murshid A, Gong J, Calderwood SK (2012) The role of heat shock proteins in antigen cross presentation. Front Immunol 3:63. Epub 2012 Mar 30. doi: 10.3389/fimmu.2012.00063

Ostan R, Bucci L, Capri M, Salvioli S, Scurti M, Pini E, Monti D, Franceschi C (2008) Immunosenescence and immunogenetics of human longevity. Neuroimmunomodulation 15:224–240

Osterloh A, Geisinger F, Piédavent M, Fleischer B, Brattig N, Breloer M (2009) Heat shock protein 60 (HSP60) stimulates neutrophil effector functions. J Leuk Biol 86:423–434

Tamura Y, Torigoe T, Kukita K, Saito K, Okuya K, Kutomi G, Hirata K, Sato N (2012) Heat-shock proteins as endogenous ligands building a bridge between innate and adaptive immunity. Immunotherapy 4:841–852

Xie J, Zhu H, Guo L, Ruan Y, Wang L, Sun L, Zhou L, Wu W, Yun X, Shen A, Gu J (2010) Lectin-like oxidized low-density lipoprotein receptor-1 delivers heat shock protein 60-fused antigen into the MHC class I presentation pathway. J Immunol 185:2306–2313

Zanin-Zhorov A, Cahalon L, Tal G, Margalit R, Lider O, Cohen IR (2006) Heat shock protein 60 enhances CD4 + CD25 + regulatory T cell function via innate TLR2 signaling. J Clin Invest 116:2022–2032

The Argument Against Chaperones Having a Hormonal-Like Effect on Cells of the Innate Immune System: The Effects Observed are Due to Contaminants

Gao B, Tsan MF (2004) Induction of cytokines by heat shock proteins and endotoxin in murine macrophages. Biochem Biophys Res Commun 317:1149–1154

Stocki P, Wang XN, Dickinson AM (2012) Inducible heat shock protein 70 reduces T cell responses and stimulatory capacity of monocyte-derived dendritic cells. J Biol Chem 287:12387–12394

Tsan MF, Gao B (2009) Heat shock proteins and immune system. J Leukoc Biol 85:905–910

Hsp60 Pure and Hsp60 Plus LPS Both are Immunostimulatory (Activate Antigen-Presenting Cells) but Via Two Separate Mechanisms

Osterloh A, Kalinke U, Weiss S, Fleischer B, Breloer M (2007) Synergistic and differential modulation of immune responses by Hsp60 and lipopolysaccharide. J Biol Chem 282:4669–4680

Osterloh A, Meier-Stiegen F, Veit A, Fleischer B, von Bonin A, Breloer M (2004) Lipopolysaccharide-free heat shock protein 60 activates T cells. J Biol Chem 279:47906–47911

Osterloh A, Veit A, Gessner A, Fleischer B, Breloer M (2008) Hsp60-mediated T cell stimulation is independent of TLR4 and IL-12. Int Immunol 20:433–443

Hsp60 and TLR4 in Post-Ischemia Myocardial Inflammation

Myocardial inflammation occurs often as a consequence of stimulation of the innate immune system following transient ischemia. It was demonstrated experimentally that myocardial ischemia activates interleukin receptor-associated kinase-1 (IRAK-1), a kinase critical for the innate immune signaling such as that of Toll-like receptors (TLRs), via a mechanism that involves Hsp60 and TLR4. This leads to cardiomyocyte apoptosis and inflammation

Li Y, Si R, Feng Y, Chen HH, Zou L, Wang E, Zhang M, Warren HS, Sosnovik DE, Chao W
    (2011b) Myocardial ischemia activates an injurious innate immune signaling via cardiac heat
    shock protein 60 and Toll-like receptor 4. J Biol Chem 286:31308–33119

**Adipocytes. Hsp60 Binds Other Molecules (in Addition to Substrate for
Folding), Cell-Surface Receptors, Cells, etc. and Induces Production of Pro-
inflammatory Cytokines and Other Factors**

Märker T, Kriebel J, Wohlrab U, Habich C (2010) Heat shock protein 60 and adipocytes:
    characterization of a ligand-receptor interaction. Biochem Biophys Res Commun 391:1634–1640
Märker T, Sell H, Zilleßen P, Glöde A, Kriebel J, Ouwens DM, Pattyn P, Ruige J, Famulla S,
    Roden M, Eckel J, Habich C (2012) Heat shock protein 60 as a mediator of adipose tissue
    inflammation and insulin resistance. Diabetes 61:615–625

**Heat Shock Protein 60: Evidence for Receptor-Mediated Induction of
Proinflammatory Mediators During Adipocyte Differentiation**

Gülden E, Märker T, Kriebel J, Kolb-Bachofen V, Burkart V, Habich C (2009) Hsp60 bind receptors
    on, and induce release of proinflammatory mediators by adipocytes. FEBS Lett 583:2877–2881

**SKIN TLR4 is More Increased in Guttate than in Plaque Psoriasis. Hsp60
Could Play a Role in Pathogenesis by Activating the Innate Immune System
Via TLR4**

Seung NR, Park EJ, Kim CW, Kim KH, Kim KJ, Cho HJ, Park HR (2007) Comparison of
    expression of heat-shock protein 60, Toll-like receptors 2 and 4, and T-cell receptor gamma
    delta in plaque and guttate psoriasis. J Cutan Pathol 34:903–911

Hsp60 Induces T cells in Atopic Dermatitis and Promotes Inflammation

Kapitein B, Aalberse JA, Klein MR, de Jager W, Hoekstra MO, Knol EF, Prakken BJ (2013)
    Recognition of self-heat shock protein 60 by T cells from patients with atopic dermatitis. Cell
    Stress Chaperones 18:87–95

CARDIOVASCULAR SYSTEM

Zal B, Kaski JC, Arno G, Akiyu JP, Xu Q, Cole D, Whelan M, Russell N, Madrigal JA, Dodi IA,
    Baboonian C (2004) Heat-shock protein 60-reactive CD4 + CD28-null T cells in patients
    with acute coronary syndromes. Circulation 109:1230–1235

HUMAN HSP60 EPITOPES INDUCE T-CELL PROLIFERATION AND PROINFLAMMATORY MEDIATORS

de Jong H, Lafeber FF, de Jager W, Haverkamp MH, Kuis W, Bijlsma JW, Prakken BJ, Albani S
    (2009) PAN-DR-Binding Hsp60 self epitopes induce an interleukin-10-mediated immune
    response in rheumatoid arthritis. Arthritis Rheum 60:1966–1976

HSP70 RELEASED FROM TUMORS ACTIVATE TUMOR CELLS TO PRODUCE CHEMOKINES FOR CHEMOATTRACTION OF DENDRITIC AND T CELLS VIA THE TLR4 SIGNALING PATHWAY

Chen T, Guo J, Han C, Yang M, Cao X (2009) Heat shock protein 70, released from heat-stressed tumor cells, initiates antitumor immunity by inducing tumor cell chemokine production and activating dendritic cells via TLR4 pathway. J Immunol 182:1449–1459

CNS Hsp60 Released from CNS Cells Activate Microglial Cells (CNS Monocytes) Via TLR4- and MYD88-Dependent Pathways

Lehnardt S, Schott E, Trimbuch T, Laubisch D, Krueger C, Wulczyn G, Nitsch R, Weber JR (2008) A vicious cycle involving release of heat shock protein 60 from injured cells and activation of toll-like receptor 4 mediates neurodegeneration in the CNS. J Neurosci 28:2320–2331

Hsp60 Interacts with Microglia Via LOX-1 and Induces Inflammation

Zhang D, Sun L, Zhu H, Wang L, Wu W, Xie J, Gu J (2012) Microglial LOX-1 reacts with extracellular HSP60 to bridge neuroinflammation and neurotoxicity. Neurochem Int 61:1021–1035

MYOPATHIES

De Paepe B, Creus KK, Martin JJ, Weis J, De Bleecker JL (2009) A dual role for HSP90 and HSP70 in the inflammatory myopathies: from muscle fiber protection to active invasion by macrophages. Ann New York Acad Sci 1173:463–469

Li H, Chen Q, Liu F, Zhang X, Li W, Liu S, Zhao Y, Gong Y, Yan C (2013) Unfolded protein response and activated degradative pathways regulation in GNE myopathy. PLoS ONE 8(3):e58116. doi:10.1371/journal.pone.0058116

Chaperones, the Unfolded Protein Response (UPR), Protein Degradation, and Autophagy in Inflammatory Myopathies, Sporadic Inclusion Body Myositis (IBM), and Polymyositis (PM)

Cacciottolo M, Nogalska A, D'Agostino C, Engel WK, Askanas V (2013) Chaperone-mediated autophagy components are upregulated in sporadic inclusion-body myositis muscle fibres. Neuropathol Appl Neurobiol. doi:10.1111/nan.12038

VMA21 (VMA21 vacuolar H+-ATPase homolog) in Myopathies

Ramachandran N, Munteanu I, Wang P, Ruggieri A, Rilstone JJ, Israelian N, Naranian T, Paroutis P, Guo R, Ren ZP, Nishino I, Chabrol B, Pellissier JF, Minetti C, Udd B, Fardeau M, Tailor CS, Mahuran DJ, Kissel JT, Kalimo H, Levy N, Manolson MF, Ackerley CA, Minassian BA (2013) VMA21 deficiency prevents vacuolar ATPase assembly and causes autophagic vacuolar myopathy. Acta Neuropathol 125:439–457. doi:10.1007/s00401-012-1073-6

RHEUMATOID ARTHRITIS ER Chaperone GRP78 is Crucial for Synoviocyte Proliferation and Angiogenesis, the Pathological Hallmark of Rheumatoid Arthritis

Yoo SA, You S, Yoon HJ, Kim DH, Kim HS, Lee K, Ahn JH, Hwang D, Lee AS, Kim KJ, Park YJ, Cho CS, Kim WU (2012) A Novel Pathogenic Role of the ER Chaperone GRP78/BiP in Rheumatoid Arthritis. J Exp Med 209:871–886

HSP60 AND BONE PATHOLOGY

Kim YS, Koh JM, Lee YS, Kim BJ, Lee SH, Lee KU, Kim GS (2009) Increased circulating heat shock protein 60 induced by menopause, stimulates apoptosis of osteoblast-lineage cells via up-regulation of toll-like receptors. Bone 45:68–76
Koh JM, Lee YS, Kim YS, Park SH, Lee SH, Kim HH, Lee MS, Lee KU, Kim GS (2009) Heat shock protein 60 causes osteoclastic bone resorption via toll-like receptor-2 in estrogen deficiency. Bone 45:650–660
Meghji S, Lillicrap M, Maguire M, Tabona P, Gaston JS, Poole S, Henderson B (2003) Human chaperonin 60 (Hsp60) stimulates bone resorption: structure/function relationships. Bone 33:419–425

MULTIPLE FUNCTIONS OF CHAPERONES BEYOND FOLDING OF NASCENT POLYPEPTIDES: CHAPERONE COMPLEXES THAT ARE NOT CHAPERONING MACHINES

Campanella C, Bucchieri F, Ardizzone NM, Marino Gammazza A, Montalbano A, Ribbene A, Di Felice V, Bellafiore M, David S, Rappa F, Marasa M, Peri G, Farina F, Czarnecka AM, Conway de Macario E, Macario AJL, Zummo G, Cappello F (2008) Upon oxidative stress, the antiapoptotic Hsp60/procaspase-3 complex persists in mucoepidermoid carcinoma cells. Eur J Histochem 52:221–228
Meimaridou E, Gooljar SB, Chapple JP (2009) From hatching to dispatching: the multiple cellular roles of the Hsp70 molecular chaperone machinery. J Mol Endocrinol 42:1–9
Smith DF, Toft DO (2008) Minireview: the intersection of steroid receptors with molecular chaperones: observations and questions. Mol Endocrinol 22:2229–2240

CHAPERONOPATHIES BY MISTAKE: A CHAPERONE ACTS AS AUTOANTIGEN

Cappello F, Marino Gammazza A, Zummo L, Conway de Macario E, Macario AJL (2010) Hsp60 and AChR cross-reactivity in myasthenia gravis: an update. J Neurol Sciences 292:117–118
Cloos PAC, Christgau S (2004) Post-translational modifications of proteins: implications for ageing, antigen recognition, and autoimmunity. Biogerontology 5:139–158

Pathogenic Anti-Hsp60 Antibodies: Cardiovascular System

Alard JE, Hillion S, Guillevin L, Saraux A, Pers JO, Youinou P, Jamin C (2011) Autoantibodies to endothelial cell surface ATP synthase, the endogenous receptor for hsp60, might play a pathogenic role in vasculatides. PLoS ONE 6(2):e14654
Grundtman C, Kreutmayer SB, Almanzar G, Wick MC, Wick G (2011) Heat shock protein 60 and immune inflammatory responses in atherosclerosis. Arterioscler Thromb Vasc Biol 31:960–968

Mayr M, Metzler B, Kiechl S, Willeit J, Schett G, Xu Q, Wick G (1999a) Endothelial cytotoxicity mediated by serum antibodies to heat shock proteins of *Escherichia coli* and *Chlamydia pneumoniae*: immune reactions to heat shock proteins as a possible link between infection and atherosclerosis. Circulation 99:1560–1566

Hsp60 and Myeloperoxidase (MPO) May Share Antigenic Determinants

If Hsp60 and MPO share antigenic determinants, antibodies to Hsp60 may cross-react with MPO and thus contribute to the pathogenesis of vasculitis associated with antineutrophil cytoplasmic antibodies (ANCA), in which MPO is a major pathogenetic autoantigen

Slot MC, Theunissen R, van Paassen P, Damoiseaux JG, Tervaert JW (2006) Evaluation of antibodies against human HSP60 in patients with MPO-ANCA associated glomerulonephritis: a cohort study. J Autoimmune Dis 3:4. doi:10.1186/1740-2557-3-4

Serum Anti-Hsp60 Antibodies Cross-React with Human Hsp60, *Chlamydia Trachomatis* or *Chlamydia Pneumoniae* Hsp60, and GroEL from other Bacteria

Serum anti-Hsp60 antibodies correlate with the presence of antibodies to *C. pneumoniae* (Cp) and endotoxin; and mediate endothelial cytotoxicity. Humoral immune reactions to bacterial Hsps, such as cpHsp60 and other bacterial GroEL, play a role in vascular endothelial injury, which is believed to be a key event in the pathogenesis of atherosclerosis

Mascellino MT, Boccia P, Oliva A (2011) Immunopathogenesis in *Chlamydia trachomatis* infectedwomen. International Scholarly Research Network ISRN Obstetrics and Gynecology Article ID 436936, 9 pages. doi: 10.5402/2011/436936

Mayr M, Metzler B, Kiechl S, Willeit J, Schett G, Xu Q, Wick G (1999b) Endothelial cytotoxicity mediated by serum antibodies to heat shock proteins of *Escherichia coli* and *Chlamydia pneumoniae*: immune reactions to heat shock proteins as a possible link between infection and atherosclerosis. Circulation 99:1560–1566

Hsp60 as Autoantigen with Pathogenic Effect in Atherosclerosis: Induces T-Cells

Rossmann A, Henderson B, Heidecker B, Seiler R, Fraedrich G, Singh M, Parson W, Keller M, Grubeck-Loebenstein B, Wick G (2008) T-cells from advanced atherosclerotic lesions recognize hHSP60 and have a restricted T-cell receptor repertoire. Exp Gerontol 43:229–237

Mitochondrial Hsp70 (mtHsp70) is the Surface Receptor of Endothelial Cell for Hsp60 and Both Interact Within Lipid Rafts

Alard JE, Dueymes M, Mageed RA, Saraux A, Youinou P, Jamin C (2009) Mitochondrial heat shock protein (HSP) 70 synergizes with HSP60 in transducing endothelial cell apoptosis induced by anti-HSP60 autoantibody. FASEB J 23:2772–2779

Hsp70 Joints Synovial

Hsp70 increases in synovial cells and localizes to the membrane and may go out of the cell and bind to the membrane on the outside of other cells

Sedlackova L, Nguyen TT, Zlacka D, Sosna A, Hromadnikova I (2009) Cell surface and relative mRNA expression of heat shock protein 70 in human synovial cells. Autoimmunity 42:17–24

Hsp60 in Autoimmunity: Cellular and Humoral Immunities

López-Hoyos M, Alvarez L, Ruiz Soto M, Blanco R, José Bartolomé M, Martínez-Taboada VM (2008) Serum levels of antibodies to *Chlamydia pneumoniae* and human HSP60 in giant cell arteritis patients. Clin Exp Rheumatol 26:1107–1110

Moudgil KD, Durai M (2008) Regulation of autoimmune arthritis by self heat-shock proteins. Trends Immunol 29:412–418

Van Roon JA, van Eden W, van Roy JL, Lafeber FP, Bijlsma JW (1997) Stimulation of suppressive T cell responses by human but not bacterial 60-kD heat-shock protein in synovial fluid of patients with rheumatoid arthritis. J Clin Invest 100:459–463

de Jong H, de Jager W, Wenting-van Wijk M, Prakken BJ, Kruize AA, Bijlsma JW, Lafeber FP, van Roon JA (2012) Increased immune reactivity towards human Hsp60 in patients with primary Sjögren's syndrome is associated with increased cytokine levels and glandular inflammation. Clin Exp Rheumatol 30:594–595

Diabetes, Obesity, Endocrine Glands

Brudzynski K (1993) Insulitis-caused redistribution of heat-shock protein HSP60 inside beta-cells correlates with induction of HSP60 autoantibodies. Diabetes 42:908–913

Mallard K, Jones DB, Richmond J, McGill M, Foulis AK (1996) Expression of the human heat shock protein 60 in thyroid, pancreatic, hepatic and adrenal autoimmunity. J Autoimmun 9:89–96

Kidney

Slot MC, Theunissen R, van Paassen P, Damoiseaux JG, Tervaert JW (2006) Evaluation of antibodies against human HSP60 in patients with MPO-ANCA associated glomerulonephritis: a cohort study. J Autoimmune Dis 5:3–4

Muscle

Elst EF, Klein M, de Jager W, Kamphuis S, Wedderburn LR, van der Zee R, Albani S, Kuis W, Prakken BJ (2008) Hsp60 in inflamed muscle tissue is the target of regulatory autoreactive T cells in patients with juvenile dermatomyositis. Arthritis Rheum 58:547–555

Skin

Seung NR, Park EJ, Kim CW, Kim KH, Kim KJ, Cho HJ, Park HR (2007) Comparison of expression of heat-shock protein 60, toll-like receptors 2 and 4, and T-cell receptor gammadelta in plaque and guttate psoriasis. J Cutan Pathol 34:903–911

Infections

Prohászka Z, Duba J, Lakos G, Kiss E, Varga L, Jánoskuti L, Császár A, Karádi I, Nagy K, Singh M, Romics L, Füst G (1999) Antibodies against human heat-shock protein (hsp) 60 and mycobacterial hsp65 differ in their antigen specificity and complement-activating ability. Int Immunol 11:1363–1370

Yamabe K, Maeda H, Kokeguchi S, Soga Y, Meguro M, Naruishi K, Asakawa S, Takashiba S (2010) Antigenic group II chaperonin in *Methanobrevibacter oralis* may cross-react with human chaperonin CCT. Mol Oral Microbiol 25:112–122

Listeria monocytogens LAP Interacts with Hsp60 in Human Cells

Jagadeesan B, Fleishman Littlejohn AE, Amalaradjou MA, Singh AK, Mishra KK, La D, Kihara D, Bhunia AK (2011) N-terminal gly (224)-gly (411) domain in listeria adhesion protein (LAP) interacts with host receptor Hsp60. PLoS ONE 6(6):e20694

Hsp60 Peptides as Modulators of Proinflammatory Cytokines

Puga Yung GL, Fidler M, Albani E, Spermon N, Teklenburg G, Newbury R, Schechter N, van den Broek T, Prakken B, Billetta R, Dohil R, Albani S (2009a) Heat shock protein-derived T-cell epitopes contribute to autoimmune inflammation in pediatric Crohn's disease. PLoS ONE 4(11):e7714

Hsp60 can Influence T-cell Response Acting as Autoantigen and as Ligand of Toll-Like Receptor 2 Signaling

Quintana FJ, Mimran A, Carmi P, Mor F, Cohen IR (2008) HSP60 as a target of anti-ergotypic regulatory T cells. PLoS ONE 3(12):e4026

Anti Hsp Autoantibodies in Newborns and in Autoimmune Diabetes and Multiple Sclerosis

Quintana FJ, Cohen IR (2008) HSP60 speaks to the immune system in many voices. Novartis Found Symp 291:101–111; discussion 111–4, 137–140

Hsp70 in Viral Infection

Lahaye X, Vidy A, Fouquet B, Blondel D (2012) Hsp70 protein positively regulates Rabies virus infection. J Virol 86:4743–4751

Hsp60: ROLE IN DISEASES WITH PATHOGENIC ACTIVATION OF THE IMMUNE SYSTEM

Presence of chaperones, Hsp60 and/or others, in affected tissues suggest their participation as a pathogenic factor, most likely by stimulating cells of the innate immune system to mount an inflammatory response

Hsp60 Chronic Obstructive Pulmonary Disease

Di Stefano A, Caramori G, Gnemmi I, Contoli M, Bristot L, Capelli A, Ricciardolo FL, Magno F, D'Anna SE, Zanini A, Carbone M, Sabatini F, Usai C, Brun P, Chung KF, Barnes PJ, Papi A, Adcock IM, Balbi B (2009) Association of increased CCL5 and CXCL7 chemokine expression with neutrophil activation in severe stable COPD. Thorax 64:968–975

Hacker S, Lambers C, Hoetzenecker K, Pollreisz A, Aigner C, Lichtenauer M, Mangold A, Niederpold T, Zimmermann M, Taghavi S, Klepetko W, Ankersmit HJ (2009) Elevated HSP27, HSP70 and HSP90 alpha in chronic obstructive pulmonary disease: markers for immune activation and tissue destruction. Clin Lab 55:31–40

Inflammatory Bowel Disease Ulcerative Colitis

Tomasello G, Sciumè C, Rappa F, Rodolico V, Zerilli M, Martorana A, Cicero G, De Luca R, Damiani P, Accardo FM, Romeo M, Farina F., Bonaventura G, Modica G, Zummo G,

Conway de Macario E, Macario AJL, Cappello F (2011) Hsp10, Hsp70, and Hsp90 immunohistochemical levels change in ulcerative colitis after therapy. Eur. J. Histochemistry 55(4) e38: 210–214, 2011. doi: 10.4081/ejh.2011.e38

Hsp60 Peptides as Modulators of Proinflammatory Cytokines: Crohn's Disease

Puga Yung GL, Fidler M, Albani E, Spermon N, Teklenburg G, Newbury R, Schechter N, van den Broek T, Prakken B, Billetta R, Dohil R, Albani S (2009b) Heat shock protein-derived T-cell epitopes contribute to autoimmune inflammation in pediatric Crohn's disease. PLoS ONE 4(11):e7714

Hsp10 Autoantigen in Pancreatitis and Serious Type 1 Diabetes

Takizawa S, Endo T, Wanjia X, Tanaka S, Takahashi M, Kobayashi T (2009) HSP 10 is a new autoantigen in both autoimmune pancreatitis and fulminant type 1 diabetes. Biochem Biophys Res Commun 386:192–196

Anti-autoinmunity (Also Tumor Promotion)

Jia H, Halilou AI, Hu L, Cai W, Liu J, Huang B (2011) Heat shock protein 10 (Hsp10) in immune-related diseases: one coin, two sides. Int J Biochem Mol Biol 2:47–57

Chapter 8
Impact of Chaperonopathies in Protein Homeostasis and Beyond

Abstract Chaperones have functions other than those classically attributed to them pertaining to protein homeostasis. These "other" non-canonical functions are the focus of chapter 8. The close interaction of the chaperoning and the immune systems and the impact of their malfunctioning on aging and cancer are highlighted. Conversely, the impact of ageing and cancer on the two systems is also underscored. The connections between stress, protein damage (including chaperones), protein misfolding, protein aggregation and precipitation, and tissue degeneration, are analyzed, indicating that all these processes are aggravated by a decline in chaperoning potential with aging (chaperonopathies of the aged) and the progression of certain pathologic conditions. Again, the potential of chaperonotherapy in these chaperonopathies is mentioned as a worthy aim in future research.

Keywords Chaperones, non-canonical functions · Chaperoning-immune system interaction · Aging · Cancer · Stress · Proteinopathy · Protein misfolding · Protein aggregation · Protein precipitates · Ubiquitin-proteasome system (UPS) · Chaperone-mediated autophagy (CMA) · Chaperonotherapy

As stated previously, a molecular chaperone can have the typical functions pertaining to protein homeostasis but it can also have other functions unrelated to protein homeostasis. It is likely that each distinct function depends to a considerable extent on where the chaperone is located and with what it associates, although subtle structural differences, e.g., posttranslational modifications may also define what a chaperone does, how, and where. Thus, chaperones overall play a variety of roles in health and disease, and these roles intersect and overlap with those of other physiological systems, for instance the immune system. Consequently, chaperones are implicated in processes such as cancer and ageing that involve more than one physiological system (Fig. 8.1).

In what regards protein homeostasis, stress and proteinopathy (generalized as in aged individuals, or restricted to certain groups of proteins as in certain congenital abnormalities for instance) have a compounded pathological effect that becomes still more devastating if one or more chaperones are defective or sick. Chaperonotherapy would, in principle be indicated (Fig. 8.2).

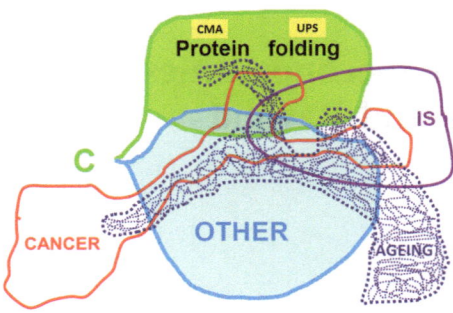

Fig. 8.1 Chaperones (*C*) have functions pertaining to protein folding and other processes relevant to maintenance of protein homeostasis, including participation in chaperone-mediated autophagy (*CMA*) and ushering proteins destined to degradation to the ubiquitin–proteasome system (*UPS*). All these functions are represented in the figure by the area named *Protein folding*. In addition, chaperones have other functions (area designated *OTHER*) only marginally related or unrelated to protein homeostasis. The latter involve, for instance, interactions with the immune system (*IS*), and playing a role in *CANCER* and *AGEING*, as indicated, with overlapping between the various functions and interactions. See also Sect. 7.2. *Source* Macario, A.J.L. and Conway de Macario E. (unpublished)

Fig. 8.2 Coexistence of a proteinopathy with stress and a chaperonopathy may seriously disrupt anti-stress mechanisms and protein homeostasis, with devasting consequences for the patient. *Source* Macario A.J.L. Conway de Macario E (2007) Chaperonopathies and chaperonotherapy. FEBS Lett 581:3681–3688

PATHOGENIC PATHWAYS
Stress >> Protein unfolding >> Aggregation
Proteinopathy >> Protein misfolding >> Aggregation
Proteinopathy + Stress = Aggregation
Chaperones to the rescue
Chaperonopathy >> Protein folding impaired
Proteinopathy + Chaperonopathy = Aggregation
Proteinopathy + Chaperonopathy + Stress = AGGREGATION

Further Reading

IMPACT OF CHAPERONOPATHIES IN PROTEIN HOMEOSTASIS AND BEYOND

For Further Reading see also Chaps. 6 and 7

Hsp60 Helps (Mistakenly?) in the Accumulation of Pathological Proteins Inside Mitochondria, Disrupting Mitochondrial Functions

Walls KC, Coskun P, Gallegos-Perez JL, Zadourian N, Freude K, Rasool S, Blurton-Jones M, Green KN, Laferla FM (2012) Swedish Alzheimer's mutation induces mitochondrial dysfunction mediated by HSP60 mislocalization of APP and beta-amyloid. J Biol Chem 287:30317–3027

Chapter 9
Extracellular Chaperones

Abstract The rapidly growing field of extracellular chaperones is the topic of chapter 9. Chaperones, once thought to reside only inside cells, are now known to also occur and work extracellularly. Many questions remain unanswered, such as when and how chaperones emerge from the subcellular compartment in which they were supposed to reside exclusively, when and how they exit the cell, and where do they go and what do they do at their destinations. To illustrate some of these hot areas of research, this chapter analyses some data available on the odyssey of Hsp60 from inside to the outside of cells in normal and some tumor cells.

Keywords Extracellular chaperones · Hsp60 secretion · Tumor cells · Cell plasma membrane · Lipid rafts · Endosomes · Multivesicular bodies · Exosomes · Golgi apparatus · Endoplasmic reticulum · Protein secretion inhibitors · Cell signaling · Circulating chaperones · Hsp cancer biomarkers · Diagnostic-prognostic indicators

The chaperoning system concept implies extracellular and circulating chaperones (see Fig. 2.5), both of which have already been demonstrated. The question is how the chaperones exit the cell. Data showing how that happens for Hsp60 in tumor cells, at least those we studied are presented in the following pages.

We found that the tumor cell lines we tested released Hsp60 into the culture medium, Fig. 9.1. The chaperone was detected in the culture medium by immunoprecipitation, and in the exosomes purified from the culture medium, suggesting that the latter are the carriers of the Hsp60 as it exits the cell.

We also demonstrated that compounds known to inhibit protein trafficking via lipid rafts and exosomes did inhibit Hsp60 secretion by the tumor cell lines without any sign of cell damage, Fig. 9.2. This finding strongly supports the notion that Hsp60 is actively secreted by viable tumor cells via a mechanism involving lipid rafts and exosomes. Subsequently, we demonstrated the presence of Hsp60 in the cell membrane of the tumor cells, and in their lipid rafts and exosomes and in the

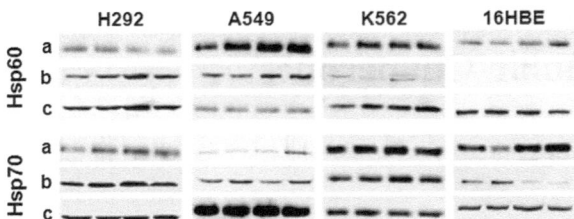

Fig. 9.1   Extracellular release of Hsp60 and Hsp70 by tumor cells. The two Hsps were found extracellularly in specific immunoprecipitates from conditioned media (**a**) and in exosomes purified from the conditioned media from all three tumor cells tested (in A549, Hsp70 levels in immunoprecipitates from conditioned media were lower than in the other cell lines) (**b**). Likewise, Hsp60 and Hsp70 were present in immunoprecipitates from conditioned medium from the 16HBE non-tumor cells. However, while Hsp70 was present in the exosomes purified from the 16HBE conditioned medium, Hsp60 was not. As expected, the two Hsps were present intracellularly in all cell lines, as shown by the results with whole-cell lysates (**c**). Each set of four Western Blot lanes represents four separate experiments. In conclusion, the two Hsps were present intracellularly and were released into the extracellular space by the tumor and non-tumor cells, but Hsp60 was not secreted via exosomes by the non-tumor cells in contrast to Hsp70. *Source* Merendino A, Bucchieri F, Campanella C, Marciano V, Ribbene A, David S, Zummo G, Burgio G, Corona D, Conway de Macario E, Macario AJL, Cappello F (2010) Hsp60 is actively secreted by human tumor cells. PLoS ONE 5(2): e9247; doi:10.1371/journal.pone.0009247, 2010. http://www.plosone.org/article/info%3Adoi%2F10.1371%2Fjournal.pone.0009247

exosomes membrane, Figs. 9.3 and 9.4. Lastly, we found evidence for involvement of the Golgi in Hsp60 trafficking, Fig. 9.5.

The data reported in Figs. 9.1, 9.2, 9.3 and 9.4, clearly point to a mechanism of active secretion of Hsp60 by tumor cells that involve the cell membrane, lipid rafts, and exosomes. Namely, the transport of Hsp60 from inside to the outside of the cell involves membranes and vesicles. In addition, we found that another trafficking modality may also participate in exporting Hsp60. We detected Hsp60 in the Golgi by direct visualization with transmission electron microscopy and immunogold labeling, Fig. 9.5a. We also found that a compound known to inhibit Golgi transport did also inhibit secretion of Hsp60 by tumor cells, Fig. 9.5b, c.

On the basis provided by all these concordant observations we propose, as a working blueprint, the Hsp60 itineraries shown in Fig. 9.6.

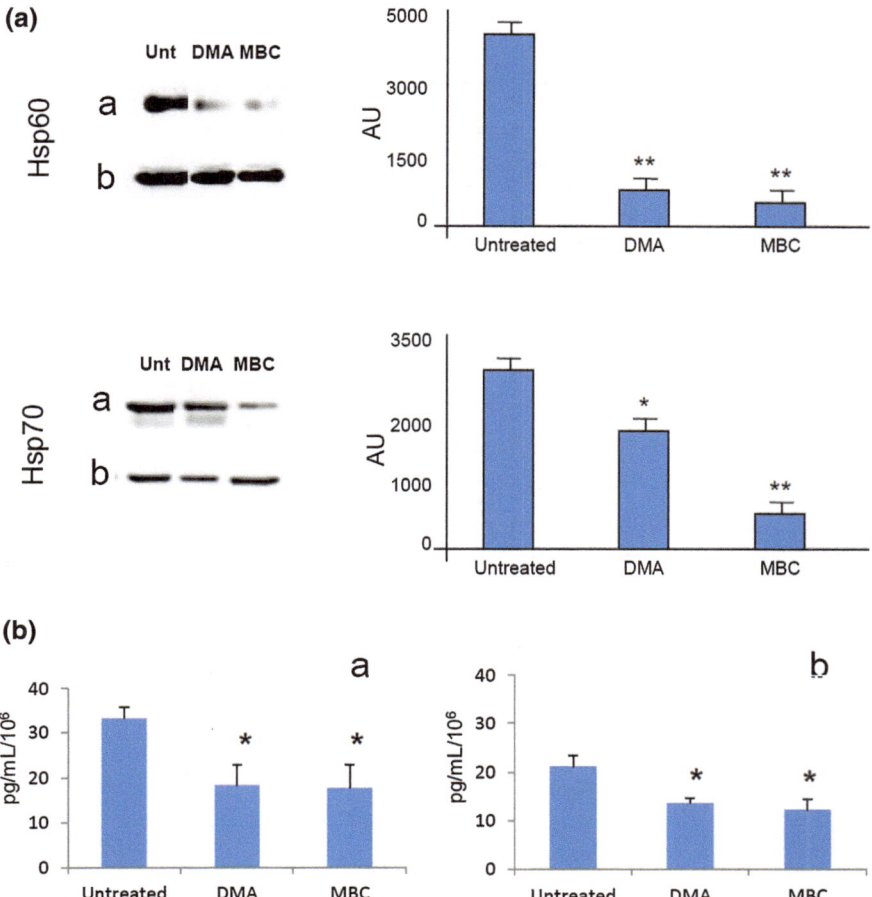

Fig. 9.2   Effects of protein-secretion inhibitors on Hsp60 secretion by tumor cells. a Hsp60 and Hsp70 detected by Western blotting in: (*a*) immunoprecipitates from conditioned media from untreated (Unt) and inhibitor-treated H292 tumor cells; and (*b*) whole-cell lysates from H292 cells. The inhibitors are listed on top of the respective lanes. Histograms to the right represent the levels of the Hsps in immunoprecipitates determined in three separate experiments as mean percentages +/2 SD of arbitrary units (AU) obtained with NIH image J 1.40 analysis software. *One asterisk* and *two asterisks*, significantly different from untreated control, p,0.005 and p,0.001, respectively. The two inhibitors (listed below the bars) significantly decreased secretion of Hsp60 and Hsp70. Also, the data from whole-cell lysates show that the protein-secretion inhibitors had no detectable effect on Hsp levels inside the cells. **b** Hsp60 levels secreted by the H292 tumor cells before and after exposure for 1 h, followed by a 4 h recovery period, to protein-secretion inhibitors measured by ELISA in: (*a*) conditioned media; and (*b*) exosomes. Histograms represent Hsp60 levels expressed as pg of protein normalised for mL in turn normalised for 10^6 cells. Data represent mean +/2 SD of three different experiments in duplicate. *Asterisk*, significantly different from untreated control, p,0.005. The results, which are in agreement with those obtained by Western blotting, show that the inhibitors tested significantly reduced secretion of Hsp60 by the H292 tumor cells. *Source* Merendino A, Bucchieri F, Campanella C, Marciano V, Ribbene A, David S, Zummo G, Burgio G, Corona D, Conway de Macario E, Macario AJL, Cappello F (2010) Hsp60 is actively secreted by human tumor cells. PLoS ONE 5(2): e9247; doi:10.1371/journal.pone.0009247, 2010. http://www.plosone.org/article/info%3Adoi%2F10.1371%2Fjournal.pone.0009247

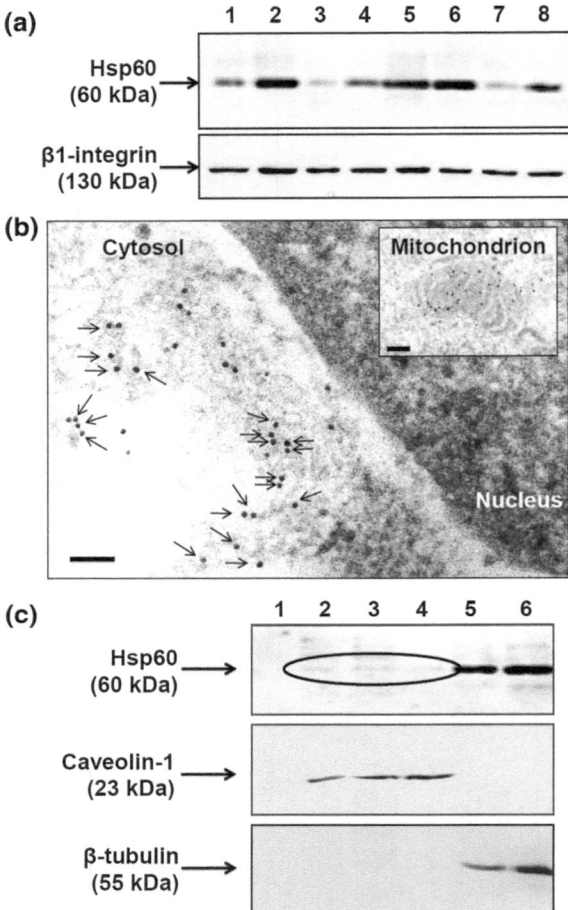

Fig. 9.3 **Hsp60 is present in the cell membrane and lipid rafts of tumor cells.** a Western blotting for Hsp60, and β1-integrin as loading control, of isolated plasma membrane. Each lane represents a separate experiment. *Lanes 1–4*, membrane from H292 tumor cells; *5–8*, membrane from A549 tumor cells. **b** Transmission electron microscopy-Immunogold demonstration of Hsp60 (*black dots*) in H292 cells. *Arrows* indicate the Hsp60 molecules that are close or onto the cell membrane. The inset shows the typical pattern of Hsp60 in a mitochondrion, which serves as a positive internal control. Bar: 100 nm. **c** Western blotting for Hsp60, caveolin-1, and β-tubulin in isolated lipid rafts from H292 tumor cells. Hsp60 is present in *lanes 2–6*; caveolin-1 (marker for lipid rafts) is present in *lanes 2–4*; and β-tubulin (cytosolic protein) is present in *lanes 5 and 6*. *Source* Campanella C, Bucchieri F, Merendino AM, Fucarino A, Burgio G, Corona DFV, Barbieri G, David, S., Farina F, Zummo G, Conway de Macario E, Macario AJL, Cappello F (2012) The Odyssey of Hsp60 from tumour cells to other destinations includes plasma membrane-associated stages and Golgi and exosomal protein-trafficking modalities. PLoS ONE 7(7): e42008; doi:10.1371/journal.pone.0042008, 2012. http://www.ncbi.nlm.nih.gov/pubmed? term=PLoS%20ONE%207(7)%3A%20e42008

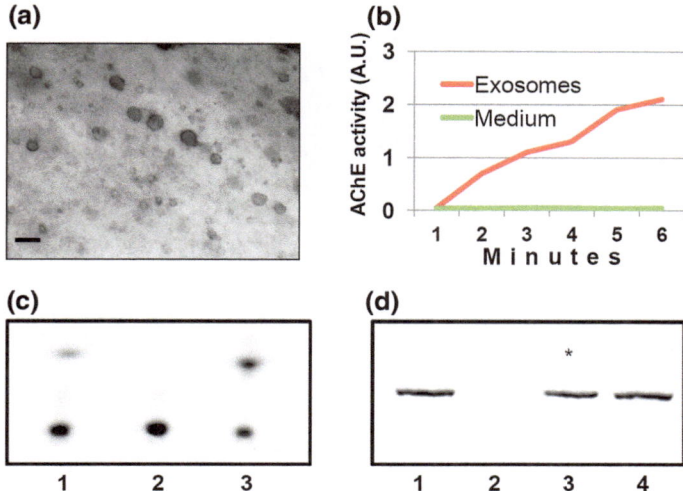

Fig. 9.4 Hsp60 is integrated in the membrane of exosomes from tumor cells. a–c Illustrative data on the exosome preparations utilized in this work. **a,** Transmission electron microscopy demonstrates that the dimension of isolated vesicles is equal to, or smaller than 100 nm, which is consistent with exosomes. Bar: 100 nm. **b** and **c,** acetylcholinesterase (AChEase) and ATPase enzymatic activities, respectively, typical of isolated exosomal vesicles, compared to control (conditioned culture medium). In b, the *red solid line* represents data from exosomes, and the *green line* represents results from conditioned culture medium. Vertical axis, AChEase activity in arbitrary units (AU), reflecting 412 nm absorbance; horizontal axis, time of reaction in minutes. In (**c**), 1, marker (positive control); 2, conditioned culture medium; 3, exosomes. **d** Treatment with sodium carbonate alone (*lane 3*, *asterisk*), or in association with proteinase K buffer (*lane 4*), does not remove Hsp60 from the exosomes, whereas treatment with Proteinase K does (*lane 2*). *Lane 1*, untreated exosomes (positive control). *Source* Campanella C, Bucchieri F, Merendino AM, Fucarino A, Burgio G, Corona DFV, Barbieri G, David S, Farina F, Zummo G, Conway de Macario E, Macario AJL, Cappello F (2012) The Odyssey of Hsp60 from tumour cells to other destinations includes plasma membrane-associated stages and Golgi and exosomal protein-trafficking modalities. PLoS ONE 7(7): e42008; doi:10.1371/journal.pone.0042008, 2012. http://www.ncbi.nlm.nih.gov/pubmed?term=PLoS%20ONE%207(7)%3A%20e42008

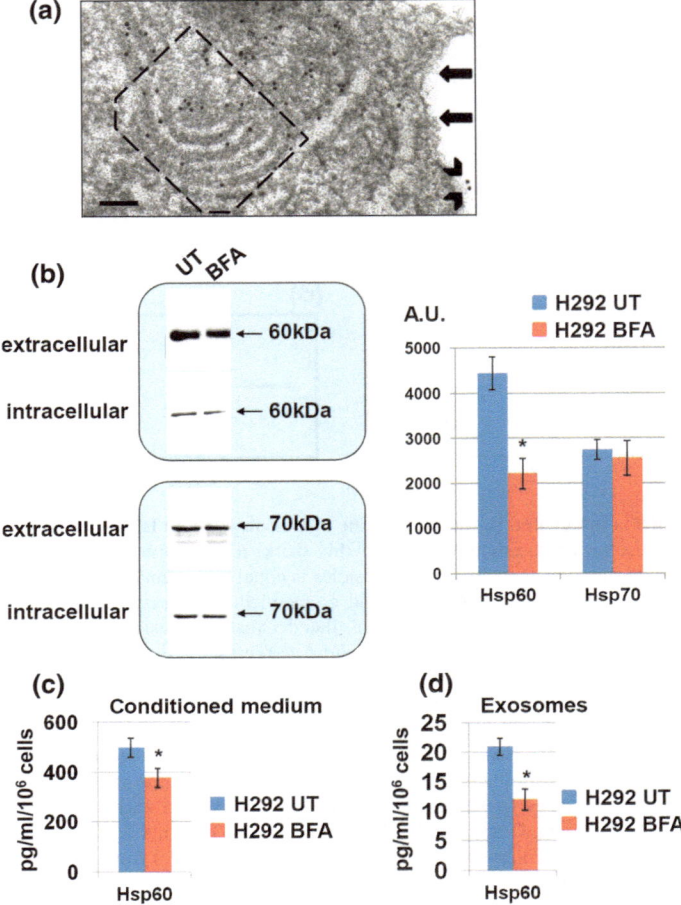

Fig. 9.5 Golgi involvement in Hsp60 secretion from tumor cells. a TEM-Immunogold shows Hsp60 (black dots) in tumor cells, including in the Golgi (framed by a dashed rectangle). The *arrows* show the plasma membrane and *arrowheads* indicate Hsp60 in it. Bar: 100 nm. **b** Extracellular levels of Hsp60 decrease after Brefeldin A (BFA) treatment of the tumor cells, as measured in immunoprecipitates from the conditioned medium (*left upper panel*). In contrast, Hsp70 levels are not influenced by BFA treatment (*left lower panel*). *Right hand panel*: histograms showing densitometric measurements of the Western blots to the left. UT, untreated; A.U.: arbitrary units; *asterisk* indicates $p < 0.005$. **c** ELISA results demonstrate a reduction of Hsp60 levels in the conditioned culture medium from tumour cells after BFA treatment. UT: untreated. **d** ELISA results show a reduction of Hsp60 levels in exosomes after BFA treatment of the tumor cells. UT: untreated. Asterisks in c and d, $p < 0.05$. Error bars represent SD. *Source* Campanella C, Bucchieri F, Merendino AM, Fucarino A, Burgio G, Corona DFV, Barbieri G, David S, Farina F, Zummo G, Conway de Macario E, Macario AJL, Cappello F (2012) The Odyssey of Hsp60 from tumor cells to other destinations includes plasma membrane-associated stages and Golgi and exosomal protein-traficking modalities. PLoS ONE 7(7): e42008; doi:10. 1371/journal.pone.0042008, 2012. http://www.ncbi.nlm.nih.gov/pubmed?term=PLoS%20ONE %207(7)%3A%20e42008

Hsp60 SECRETION FROM TUMOR CELLS AND VOYAGES TO OTHER DESTINATIONS

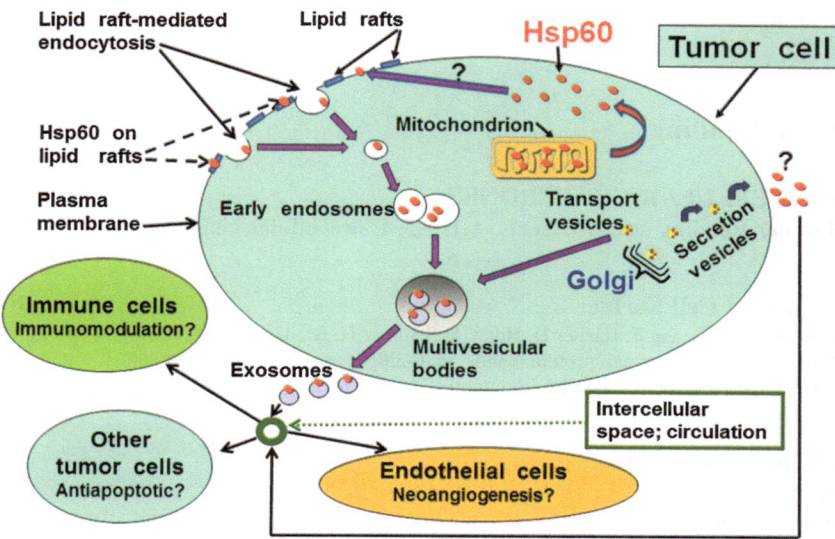

Fig. 9.6 Proposed Hsp60 secretion pathways in a tumor cell and migration to other destinations. Hsp60 (*red dots*) that in normal cells localizes mainly in mitochondria, in tumor cells accumulates also in cytosol and, for unknown reasons (post-translational modifications?), reaches the cell membrane and the Golgi. At the membrane, lipid rafts internalize (endocytose) Hsp60 toward multivesicular bodies from where it is secreted via exosomes. In these, it is located in the membrane and probably also inside. Hsp60-loaded exosomes thus reach other cells near and far through the circulation. The Golgi also participates in Hsp60 secretion via transport vesicles moving to both the multivesicular bodies and the extracellular space. Hsp60 released in the extracellular space by Golgi vesicles (free Hsp60) can thus reach other cells in the vicinity and distant via circulation. See also Sect. 2.5. *Source* Rizzo M, Macario AJL, Conway de Macario E, Gouni-Berthold I, Berthold H, Rini GB, Zummo G, Cappello F (2011) Heat-shock protein 60 and risk for cardiovascular disease. Curr Pharmaceutical Design. 17:3662–3668; and Campanella C, Bucchieri F, Merendino AM, Fucarino A, Burgio G, Corona DFV, Barbieri G, David S, Farina F, Zummo G, Conway de Macario E, Macario AJL, Cappello F (2012) The Odyssey of Hsp60 from tumour cells to other destinations includes plasma membrane-associated stages and Golgi and exosomal protein-trafficking modalities. PLoS ONE 7(7): e42008; doi:10.1371/journal.pone.0042008, 2012. http://www.ncbi.nlm.nih.gov/pubmed?term=PLoS%20ONE%207(7)%3A%20e42008

9.1 Perspective

Although the occurrence of extracellular chaperones and the mechanism of secretion by some tumor cells are established, there are various issues still unclear. For example, does secretion of Hsp60 by tumor cells, as described above for cell lines in vitro occur also in vivo, in patients with cancer? Are the tumor cells in vivo secreting Hsp60 via the same mechanisms observed in vitro, namely via exosomes and the Golgi? Preliminary information from the authors' laboratories indicate that the same occurs in vivo: patients with colon cancer have high levels of exosomes with Hsp60 in their blood but these levels return to normal after surgical ablation of the tumor.

Further Reading

EXTRACELLULAR CHAPERONES
For Further Reading see also Sects. 1.1–1.5, Extracellular chaperones

Hsp60 on the Cell Surface
Rahlff J, Trusch M, Haag F, Bacher U, Horst A, Schlüter H, Binder M (2012) Antigen-specificity of oligo clonal abnormal protein bands in multiple myeloma after allogeneic stem cell transplantation. Cancer Immunol Immunother 61:1639–1651

Hsp60 Secretion Via Exosomes
Lv LH, Wan YL, Lin Y, Zhang W, Yang M, Li GL, Lin HM, Shang CZ, Chen YJ, Min J (2012) Anticancer drugs cause release of exosomes with heat shock proteins from human hepatocellular carcinoma cells that elicit effective natural killer cell anti-tumor responses in vitro. J Biol Chem 287:15874–15885

Hsp60 and Apoptosis
Chandra D, Choy G, Tang DG (2007) Cytosolic accumulation of HSP60 during apoptosis with or without apparent mitochondrial release: evidence that its pro-apoptotic or pro-survival functions involve differential interactions with caspase-3. J Biol Chem 282:31289–31301

Hsp60 and Exosomes in the Cardiovascular System; Apoptosis
Gupta S, Knowlton AA (2005) HSP60, Bax, apoptosis and the heart. J Cell Mol Med 9:51–58
Gupta S, Knowlton AA (2007) HSP60 trafficking in adult cardiac myocytes: role of the exosomal pathway. Am J Physiol Heart Circ Physiol 292:H3052–H3056

Extracellular Hsp70
Wheeler DS, Chase MA, Senft AP, Poynter SE, Wong HR, Page K (2009) Extracellular Hsp72, an endogenous DAMP, is released by virally infected airway epithelial cells and activates neutrophils via Toll-like receptor (TLR)-4. Respir Res 10:31